T0327490

Coloring
the Cosmetic World

Coloring
the Cosmetic World

Using Pigments in Decorative
Cosmetic Formulations

Edwin B. Faulkner

Retired Director – Cosmetics, Sun Chemical

Edited by
Jane C. Hollenberg

JCH Consulting
Red Hook, NY, USA

SECOND EDITION

WILEY

Edition History
Allured Publishing Corporation (1e 2012)

Registered Offices
John Wiley & Sons, Inc., 111 River Street, Hoboken, NJ 07030, USA
John Wiley & Sons Ltd, The Atrium, Southern Gate, Chichester, West Sussex, PO19 8SQ, UK

Editorial Office
The Atrium, Southern Gate, Chichester, West Sussex, PO19 8SQ, UK

For details of our global editorial offices, customer services, and more information about Wiley products visit us at www.wiley.com.

Wiley also publishes its books in a variety of electronic formats and by print-on-demand. Some content that appears in standard print versions of this book may not be available in other formats.

Library of Congress Cataloging-in-Publication Data

Names: Faulkner, Edwin B., author. | Hollenberg, Jane C., editor.
Title: Coloring the cosmetic world : using pigments in decorative cosmetic
    formulations / Edwin B. Faulkner ; edited by Jane C. Hollenberg.
Description: Second edition. | Hoboken, NJ : Wiley, 2021. | Includes
    bibliographical references and index.
Identifiers: LCCN 2020031693 (print) | LCCN 2020031694 (ebook) | ISBN
    9781119558101 (cloth) | ISBN 9781119558118 (adobe pdf) | ISBN
    9781119558132 (epub)
Subjects: LCSH: Coloring matter. | Cosmetics.
Classification: LCC TP910 .F38 2021 (print) | LCC TP910 (ebook) | DDC
    668/.55–dc23
LC record available at https://lccn.loc.gov/2020031693
LC ebook record available at https://lccn.loc.gov/2020031694

Cover Design: Wiley
Cover Image: agsandrew/iStock/Getty Images

Set in 11.5/14pt STIXTwoText by Straive, Chennai, India

10 9 8 7 6 5 4 3 2 1

Contents

Author's Biography

E d Faulkner graduated with a degree in chemistry from Widener University and spent 44 years in the color industry. He retired from Sun Chemical Corporation in September of 2014, where his last assignment was as General Manager – Global Cosmetics and Personal Care. During his 41 year career with the company, Ed held numerous technical, manufacturing (including stints as plant manager at factories in Staten Island, NY and Cincinnati, OH), administrative, financial, sales, regulatory, and general management positions. His responsibilities necessitated a travel schedule that included North America, South America, Europe, and Asia (~35 countries and 100 cities).

For more than 20 years, Ed was a member of the adjunct faculty at the University of Cincinnati, teaching color chemistry in the Cosmetic Science Masters program. He has had numerous articles published on organic pigments, and authored the Color Cosmetic Additives chapter in *The Chemistry and Manufacture of Cosmetics* (Allured, 2009). He served as co-editor of the 2nd edition of *High Performance Pigments* (Wiley, 2009). He has lectured on the subject of pigments in many parts of the world, including the United States, United Kingdom, France, Venezuela, Chile, Japan, the Netherlands, and China.

About the Editor

Jane Hollenberg has over 40 years' experience in the cosmetic industry, working with fillers, pigments, and color cosmetics at Coty, Revlon, and Rona. Since 1996, she has operated JCH Consulting to provide services in the formulation, scale up, and troubleshooting of pigmented cosmetics.

Jane has taught in the education programs of the US Society of Cosmetic Chemists (with Edwin Faulkner) and Fairleigh Dickenson University's Masters in Cosmetic Science program, and succeeded Edwin Faulkner teaching color in the University of Cincinnati's Masters program in Cosmetic Science. She has given many lectures at industry meetings and symposia on topics relating to cosmetic pigments and colored cosmetic formulations and holds a number of patents pertaining to pigments and pigmented formulations. She authored *Color Cosmetics: A Practical Guide to Formulation* (Allured, 2016) and wrote chapters pertaining to color cosmetics and cosmetic pigments for *Chemistry & Manufacture of Cosmetics* (Allured, 2009), *Harry's Cosmeticology* (Chemical Publishing Co., 2000), and *Surfactants in Personal Care Products and Decorative Cosmetics* (CRC Press, 2006).

Preface

It has been almost 50 years since I graduated from college and entered the world of color. Like most of my colleagues in the industry, I got into color by accident. It was the only job available to me during the 1971 recession that was even remotely related to my degree in chemistry.

I spent the first 15 years of the color life in manufacturing. During that time, I became quite familiar with the chemistry and technology of the production of colors, both dyes and pigments. In 1986, I moved to the commercial side of the business.

Very little of my manufacturing background was related to the use of color in its final application areas. Therefore, I quickly found that I had a lot to learn. I started my end use education by searching for reference books on cosmetic colors and their applications. To my surprise and disappointment, I found that very little material on the use of color in cosmetic products had been written down.

Fortunately, I had my business colleague and good friend, Joe Pisetzner, for a teacher and was, therefore, able to learn from one of the masters. I dedicate this book to him, because, without Joe's patience and willingness to share his fifty years' worth of experience with me, I would not have acquired the basic knowledge to write it. The industry is very lucky to have had people like Joe at Sun Chemical and Sam Zuckerman at H.K. Kohnstamm and Aaron Cohen at Clark Colors to develop modern cosmetic color technology.

All of this brings me to the reason for writing this book. I know that there are always new cosmetic chemists coming into the industry who would like to have a reference book on color use, but, like me in my first years in the business, can't find one.

I hope that this book will fill the need that exists for these new cosmetic chemists in the industry. It contains information on the basic topics involved with the use of color in cosmetics and toiletries. It does not contain a large amount of technical details on the chemistry of the colorants themselves, but rather, it concentrates more on the practical use of them in the daily life of a cosmetic chemist.

Edwin B. Faulkner
February 2021

Editor: I, in turn, dedicate this updated second edition to my friend and collaborator, the author Ed Faulkner, who continued our mentors' practice of engaging and educating the users of cosmetic pigments.

Jane C. Hollenberg
February 2021

Acknowledgments

The author and the editor owe a great deal of thanks to a number of people and companies who helped with the preparation and the update of this book, and we would like to acknowledge them here.

James So, Sun Chemical Corporation, for his update of the regulatory information in Chapter 2.

Judith Pharo, Sun Chemical Corporation, for the preparation of many of the displays in Chapter 7.

Sun Chemical Corporation for the numerous images and support provided.

Carol Edridge, Color Techniques, Inc., for technical support and images.

Kurt Burmeister, Sensient Cosmetic Technologies, for samples and technical data on natural colors.

Paloma Moya del Valle, Eckart, for the images and information on metallic and effect pigments.

Naomi Richfield-Fratz, Beth Meyers, and Cindy Lachin, US FDA, for help with historical images and the history of the FDA.

Anita Curry, Personal Care Products Council, for her help in obtaining old Federal Register documents, and, more importantly, for all the help and knowledge she shared with us over 30 years.

Karen Willand, X-Rite Incorporated, for providing images of color space and attributes of light.

Tom DiPietro, DayGlo Corporation, for providing data, samples, and
information on fluorescent pigments.

Mark Lombardi, Konica-Minolta, Inc., for providing images of color
space and color representation.

Fred LaFaso, American Glitters, Inc., for the data, samples, and education
provided on glitter pigments.

Linbraze Ltd. for the data provided on silver pigments.

Christian Scheuring, J.G. Eytzinger GmbH, for the data provided on
gold pigments.

Gabriel Uzunian, BASF, for technical information and support on effect
pigments

Introduction

T he use of color additives in cosmetic products to change, enhance, or improve personal appearance can be traced back to antiquity. The most notable use of this technique can be found in ancient Egypt. The Egyptians used several naturally occurring colorants to produce color cosmetics, such as Red Ochre (iron), Malachite (copper), and Galena (lead). Some of these materials would cause much consternation today because of their toxic nature. Fortunately, the colorants used in today's decorative cosmetic products must meet strict criteria for use in all major world markets. This book will concentrate on the selection of colorants for cosmetic products in today's marketplace, based on a much better scientific understanding of the safety impact of chemicals on human physiology.

The selection of color additives for use in cosmetic products can be divided into four distinct criteria, each building upon its predecessor. The first is regulations. In order for a color additive to be acceptable for use in a cosmetic product, it must meet the regulatory requirements of the country where the product will be sold. No matter how "good" it is, a color additive is useless if it is not permitted by a country's regulations.

Once it is determined that a colorant meets the regulatory requirements of a particular country, the next area to consider is stability, both chemical and physical. In order to be useful, a colorant must not interact negatively with the other chemicals in the formula, the conditions of manufacture, the finished goods package, or the environment to which the final cosmetic product will be exposed. Some of the parameters to be considered with regard to stability are other chemicals, packaging materials, heat, light, pH, and humidity.

Third, once the regulatory and stability aspects are satisfied, color selection based on the desired color esthetics can be considered. This is where the cosmetic chemist can use his or her creativity to create unique color shades dictated by current fashion trends and demands. The fourth criterion relates to the third one and is economics. In today's very competitive market, there is always pressure on formulators to develop formulas in the most economical way, and the selection of colorants used can have a dramatic impact on the overall cost of the formula.

Chapters 2 through 5 are each devoted, respectively, to taking a deeper look at each of these criteria. The subsequent chapters deal with other aspects of color usage, along with discussions about effect and specialty pigments.

There are dozens of colorants approved for use in cosmetic and toiletry products in the United States and in other major world markets. They come in all shapes and sizes. Some are organic, some inorganic. Some natural, some synthetic. However, as a practical matter, only a small number of them are actually used on a regular basis. Regulatory restrictions, stability issues, color performance, and economics narrow the field of those used considerably. In order to focus attention on the color additives that are most frequently used, the topics covered in the chapters of this book will focus specifically on those colorants.

Chapter 1
Color Basics

It's a Colorful World

Is color a necessity of life, in the same manner as food, water, and air? The answer is no, but try to imagine a world without color. Nature would bear no brightly colored flowers, no green grass, no festively colored birds or fish. The blue of sky – rich and deep one day, only to yield a softer sky blue the next – would not be there. How could we call it sky without those colors? What would the word, the idea, "sky" even mean? There would also be no colorful manmade objects: automobiles gleaming red, silver, piercing black; brightly colored newspapers and magazines whose purpose is to give image and substance to ideas, concepts, and actions; children's toys and all that is implied by the joy given by their colorfulness; packaging materials that excite and endorse; the very houses we live in; and countless other objects that make up the world as we know it.

Even though color is not counted as one of life's necessities, it is an integral part of every human being's existence. It is a fact, beyond contestation, that there is color. Color pervades human life on two levels: the visual and the psychological. Visually, it is a welcome companion to human life. It provides texture, differentiation, and emotional value. In addition to the esthetic enhancement of life, color is functional in several aspects. It is used to communicate information. Think of a traffic signal. No matter in which part of the world a person may travel, he or she can, without speaking the local language, understand what the red, green, and yellow lights mean when they are illuminated, and act according to that meaning.

Coloring the Cosmetic World: Using Pigments in Decorative Cosmetic Formulations,
Second Edition. Edwin B. Faulkner. Edited by Jane C. Hollenberg.
© 2021 John Wiley & Sons Ltd. Published 2021 by John Wiley & Sons Ltd.

Along these same lines, it is possible to determine the terrorist alert level in the United States simply by seeing what color it is. Color, in this case, corresponds to a set of criteria, upon which we may become informed or, if necessary, act.

A fine example of color communication can be found in the American National Standards Institute (ANSI) color coding system (http://www.ansi.org), which can be found in Table 1.1. This system not only assigns a valuation to color, as it is used in the public sphere of everyday American life, but also offers common examples of color function in places such as public utilities, traffic coordination, and general public safety.

Beyond these functional aspects, color brings other dimensions to everyday lives. First, it is a very powerful tool for use in brand identity. There are many large multinational companies that use color as an integral part of their identity, as the following examples will attest. Coca-Cola has used red as an identifier for many years; in fact, the Coca-Cola brand is so deeply tied to this color that one knows the brand identity simply from seeing this rich red accented by the silver wave, regardless of whether the word "Coca-Cola" is present or not. Royal Dutch Shell selected yellow as its brand color and has used it very effectively in its seashell-esque logo. The United Parcel Service not only uses brown as its identifier visually, but has also incorporated the word "brown" into its corporate tag line, "What can brown do for you today?"

Along the same identity lines, companies also use color to brand specific items within their product portfolios. Over time, the colors chosen to represent a given brand, through exposure and consistent marketing plans, become practically inextricable from the product itself. To go back to Coca-Cola, no one picks up a Coke and expects to taste Pepsi, just as no one develops a taste for Coke and then looks for the Pepsi label's color scheme. Other examples of this type of color association are the bright, clean-looking orange trimmed in deep blue and white of Procter & Gamble's Tide, but a white background and the text in blue and orange for the fragrance-free version, or Cadbury Creme Eggs' distinct shiny purple wrapper.

Moving from brand and product identification, color usage is also very much a part of the foods we eat, both overtly and covertly. With regard to the former, each of us, dating back to childhood, has been instructed to "eat our greens," the idea being that green foods bear healthful nutrients beneficial to growth and development. Color is also a very prominent factor in prepared foods such as candies (the chocolatey brown wrapper of M&Ms),

Table 1.1   American National Standards Institute color coding system.

Color	Std	Meaning	Example
Red	APWA	Electric power lines, cables, conduit and lighting cables	Buried high-voltage line
Yellow	APWA	Gas, oil, steam, petroleum or gaseous materials	Natural gas distribution line
Orange	APWA	Communication, alarm or signal lines, cables or conduit	Buried fiber optic cable
Blue	APWA	Water, irrigation and slurry lines	Buried water line
Green	APWA	Sewers, drain line	Buried storm sewer line
Red	Traffic	Prohibition	STOP, yield
Orange	Traffic	Temporary work zone	Construction zone ahead
Yellow	Traffic	Caution	Merge ahead, slow
Blue	Traffic	Information signs	Hospital
Brown	Traffic	Historical or park signs	Picnic area ahead
Green	Traffic	Directional signs	Exit 1 mile, go
Red	ANSI sign	Danger signs, highest hazard	Hazardous voltage will cause death
Orange	ANSI sign	Warning signs, medium hazard	Hazardous voltage may cause death
Yellow	ANSI sign	Caution signs, lowest hazard	Turn machine off when not in use
Blue	ANSI sign	Notice signs	Employees only
Green	ANSI sign	Safety first signs	Wear ear plugs
Red	ANSI pipe	Fire quenching materials	Fire protection water
Yellow	ANSI pipe	Materials inherently hazardous	Chlorine
Blue	ANSI pipe	Materials of inherently low hazard, gas	Compressed air
Green	ANSI pipe	Materials of inherently low hazard, liquid	Storm drain
Blue/ Red/ Yellow	NFPA 701	Blue is used for health hazards, red for flammability and yellow for reactivity	

bottled and powder-mix drinks (the fun brightness of the Kool-Aid Man), and cereals (the "rainbow" of flavors suggested by Lucky Charms). As for the covert role of color, it is used to enhance the appearance of foods from raw beef to McDonald's golden fried Chicken McNuggets.

One final aspect of visual color use in everyday life is very important to any reader of this book. It is the essential core of the fashion industry, of which decorative cosmetics are a key element. Fashion, whether home, apparel, or other types, would be a three-legged stool missing a leg if everything were black and white.

The second level of color pervasion in human existence is psychological. Colors mean different things on this level than on the visual level, and can significantly affect human moods and attitudes toward events and people. Specifics of this phenomenon are as follows (http://www.infoplease.com/spot/colors1.html):

Black: Tends to signify power and authority. Ever heard someone associate the term "power" with the color black, for example in an office setting wherein one may wear a "black power suit"? It is used in fashion so pervasively because there is a belief that it makes people look thinner. Black is often used to signify villainy and darkness of character; think Darth Vader, Dracula, or Batman. On the other end of the spectrum, priests wear black to show submission to God, and, of course, black is known as the color of mourning in the Western world.

White: The color of purity and innocence. It is commonly worn by brides in the Western hemisphere (though, interestingly, it is a color of mourning in certain parts of Asia). It is often used in summer fashions, as it reflects light and is therefore cooler in hot weather. White's popularity in home fashions and clothing is largely due to the fact that it goes with all other colors.

Red: Heats up the emotions. Commonly the color associated with love or matters of the heart – think Valentine's Day. Red clothing has the opposite effect of black with respect to a person's weight appearance and is not considered conducive as an apparel choice for successful negotiations.

Blue: Red's counterpoint, it evokes calmness and tranquility; think of the soothing effect a blue sky or gently rippling blue ocean has on one's feelings of well-being. Due to this, it is one of the most popular colors and is commonly used in bedrooms and other places where a tranquil effect is desired.

Green: The corollary color to blue, as it produces a similar calm in one's emotions, as evidenced by the reference to a "green room" where people can relax while waiting to go on to a television set. For the same reason, it is a common color for hospital scrubs worn by doctors and nurses, and, in fact, hospital rooms themselves often sport walls painted a soothing shade of green. It is the color of nature and is therefore, in today's world, used to symbolize what is good for the earth. It is not much of a surprise that there is no environmental "Blue Movement."

Yellow: The color associated with the sun, making it one that promotes cheerfulness in people. However, it has been shown that too much yellow in a room can increase the chance of people losing their tempers and can even cause babies to cry more frequently. Yellow is thought to focus attention, so it is often the color of choice in legal pads.

Purple: The color of royalty, symbolizing wealth and sophistication, and also of passion, be it artistic, emotional, or spiritual. It is commonly used in the vestments and decorations of Christian churches throughout the world to denote the sovereignty of God and the "passion" by which He rose to heaven and by which the faithful show their allegiance.

Brown: The color of earth (i.e., ground, dirt, soil) and, like earth, carries the connotation of stability and reliability. It is commonly selected by men as a favorite color.

Color Theory and Color Space

Before embarking on a discussion of color selection, it is important to build a knowledge base on the fundamentals of simple color theory, color physics, human physiology with respect to color perception, color chemistry, color properties, and color nomenclature. Throughout this chapter, the terms *color additive* and *colorant* are frequently used. These are synonymous and refer to the broad class of chemicals used to impart color effects to cosmetics and toiletry products; therefore, please regard them in this manner.

As a starting point in understanding color, it is necessary to go back more than 100 years to 1905, when a scientist named Alfred Munsell postulated a theory of color that was to become the basis for the fundamental principles still in use today to explain and measure color. Munsell's theory of color space specified three attributes or dimensions of color: hue, brightness, and intensity. The first, hue, is what would commonly be called shade or, simply, color, as in red, blue, green, yellow, and so forth. This attribute is not as simple as it first sounds, because there are no pure color shades, merely degrees of differentiation. For example, a red can be a yellow shade red or a blue shade red. Likewise, a blue can be a green shade blue or a red shade blue. This pattern follows through all the colors of the visible spectrum. The second color attribute, brightness, also referred to as value, describes how dark or light a color is. Finally, intensity is the measure of a

Figure 1.1   Munsell's Color Wheel.
Source: courtesy of Konica-Minolta.

color's saturation (i.e. strong color vs. weak color), also synonymously called strength, vividness, chroma, and tinctorial value.

Once his attributes were developed and defined, Munsell went on to explore and develop the relationships among them and how these could be portrayed in a three-dimensional space, for which he created the "Color Wheel," shown in three-dimensional outline form in Figure 1.1 to give the reader an immediate sense of orientation and proportion, along with a better picture of how the color space works. *Hue* is represented by the change in shade around the horizontal plane of the wheel. *Brightness* is represented by the degree of lightness/darkness in the vertical plane. *Intensity* is represented by an increase in strength from the center of the wheel to the outer spokes. Figure 1.2 shows the red section of the same outline image rendered three dimensionally in color, demonstrating how the full spectrum is oriented. Please note that the color relationships exist on both vertical and horizontal axes of the wheel, suggesting that the assignment of Munsell's three attributes effectively both quantifies and qualifies color gradation. The subject of color space will be explored in much more depth in Chapter 7.

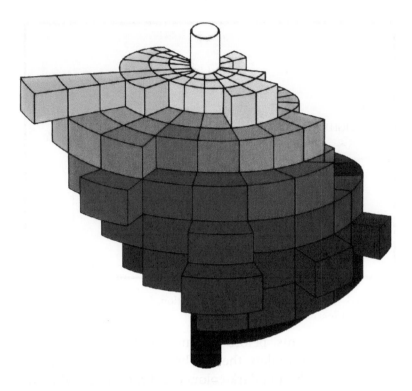

Figure 1.2 Color Wheel in 3D.
Source: courtesy of Konica-Minolta.

Human Color Physiology

The eye is a marvelous instrument. And in terms of this discussion of color, it is in truth the most important part of the human body. It facilitates the absorption and enjoyment of the vast world of color. A rendering of the basic requirements for color perception can be found in Figure 1.3. Understanding how the eye works in conjunction with the brain is fundamental to understanding how color perception and measurement work in today's sophisticated world of spectrophotometers and computers. The human eye acts in the same way as a spectrophotometer, an instrument that measures light intensity; the brain fills the role of the computer by providing the mathematical analysis. The relationship between the spectrophotometer and the computer is explained fully in Chapter 7.

Light enters the eye through the cornea and is focused by the lens on to the retina, which contains cells, called rods and cones, that separate

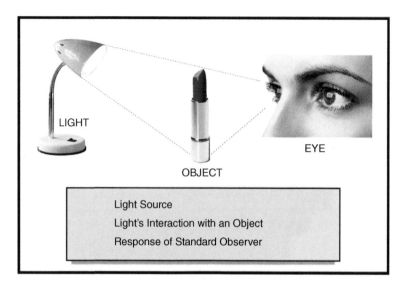

Figure 1.3   Requirements for color perception.
Source: Courtesy of Sun Chemical Corp.

color into its components, before transmitting the data to the brain for interpretation. The rods allow the eye/brain connection to perceive the difference between light and dark colors (i.e. brightness), while the cones separate color into its components of hue and judge its intensity. Hue is separated into two basic hue planes: the blue-yellow plane and the red-green plane. These color planes and light/dark perceptions are represented in Figure 1.4.

Understanding this physiology is important for two reasons. First, it helps to explain the phenomenon of color blindness in humans. There is no such thing as a person who only sees black and white. This is a common misconception passively believed by those who are *not* themselves colorblind. Color blindness is caused by defective cones in one of the two color planes, so that colorblind people either cannot distinguish between red and green or cannot distinguish between blue and yellow, but seldom both. Second, it helps to reinforce Munsell's attributes of color and explain how modern color measurement works. (Again, Chapter 7 will delve into this further.)

Color Physics

All of the color additives discussed in this chapter fall into the broad category of absorption colorants. That is, they produce their color shades through

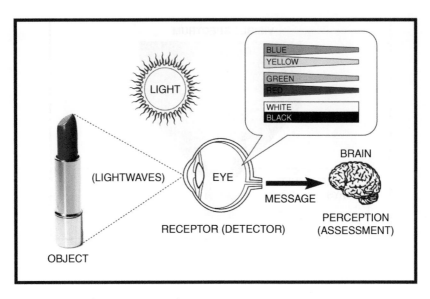

Figure 1.4   Color perception schematic.
Source: Courtesy of Sun Chemical Corp.

selective absorption and reflection of visible light. Effect pigments, which produce their color by light interference, will be examined in Chapter 9. In order for color to be perceived and evaluated, three elements are necessary. The first is a source of light, either natural (sunlight) or artificial (fluorescent, incandescent, etc.). The second is an object for which color perception is desired. The third is an observer or receptor, commonly the human eye, or, alternatively, a color measurement instrument.

As can be seen in Figure 1.5, visible light is the narrow portion of the electromagnetic spectrum that can be seen by the human eye. The band is in the area with wavelengths between 400 and 700 nanometers (nm), as shown in Figure 1.6. The lower end, near 400 nm, is the blue area, and the upper end, near 700 nm, is the red area. White light, the reader will notice, is that light which contains all of the wavelengths in the visible spectrum band.

When color additives are incorporated into cosmetic products and then bombarded with white light, they absorb some of the wavelengths of that light and reflect others. Those reflected are the ones that are perceived by the human eye as the color associated with the product. For example, the lipstick shown in Figure 1.7 absorbs the wavelengths in the 400–600 nm area and reflects those in the 600–700 nm area, thus producing the lipstick's perceived red coloration. Figure 1.8 shows how the color of this lipstick would

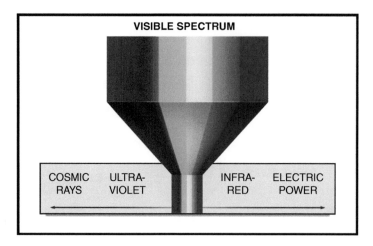

Figure 1.5   Electromagnetic spectrum.
Source: courtesy of Sun Chemical Corp.

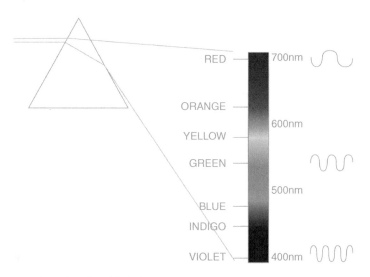

Figure 1.6   Wavelengths of visible light.
Source: © X-Rite Incorporated I www.xrite.com.

be represented by spectrophotometric light reflection measurement across the visible light range. This reflection allows the observer's eye to see the red of the lipstick. A white color is one produced when an object reflects all of the wavelengths between 400 and 700 nm, while a black color is one that absorbs all of those wavelengths.

Figure 1.7 Light reflection by a red lipstick.
Source: © X-Rite Incorporated I www.xrite.com.

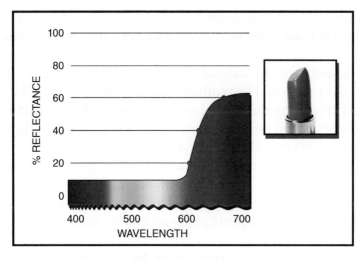

Figure 1.8 Spectrum of Figure 1.7 light absorption.
Source: © X-Rite Incorporated I www.xrite.com.

Color Chemistry

Color additives fall into one of two broad categories: dyes and pigments. *Dyes* are those color additives that are fully soluble in the medium in which

Table 1.2   Dyes and pigments.

Dyes	Pigments
Soluble	Insoluble
Color is transparent	Color films are opaque
Dissolved to impart color	Color imparted through dispersion
Pure dye content >90%	Pure dye content 10–50%
Not particulate in use	Crystalline in use

they are used, while *pigments* are those that are insoluble in the medium in which they are used. As a general rule in cosmetic product development, the term "solubility" relates to water solubility, or an additive's ability to dissolve into water, as opposed to oil or another base medium. In addition to solubility, the other major difference between the two types of colorants is that dyes are transparent and pigments are opaque. The majority of the color additives used in toiletries are dyes, whereas the majority of those used in decorative cosmetics are pigments. The reason for this is that most colored toiletry products, such as shampoos, mouthwashes, gel deodorants, and clear toothpastes, are transparent and therefore require the transparency of dyes in order to exhibit their color. Color cosmetics, on the other hand, are designed to cover the skin, so they benefit from the opacity exhibited by pigments.

Both dyes and pigments can either be natural or synthetic, and both find use in cosmetic products. The synthetic colorants normally exhibit brighter, more intense color than the natural ones, and generally better stability with respect to chemical and physical interactions. Table 1.2 shows the complete list of property differences between dyes and pigments.

Pigments are divided into two broad categories: organic and inorganic, both types being used in cosmetic products. The organic pigments are, generally speaking, brighter and more intense than the inorganic ones. On the other hand, the inorganic pigments exhibit better stability than the organic ones.

Organic pigments fall into three classes, based on the colorants' individual chemistry with respect to insolubility, as follows:

Lakes: Pigments made by absorbing a water-soluble dye on to an insoluble, inorganic substrate. There is no chemical bond between the dye and the substrate; the dye simply takes on the insoluble nature of the substrate, thereby creating a pigment. Typical substrates are aluminum hydrate and aluminum benzoate. Based on the absorptive powers of the substrate,

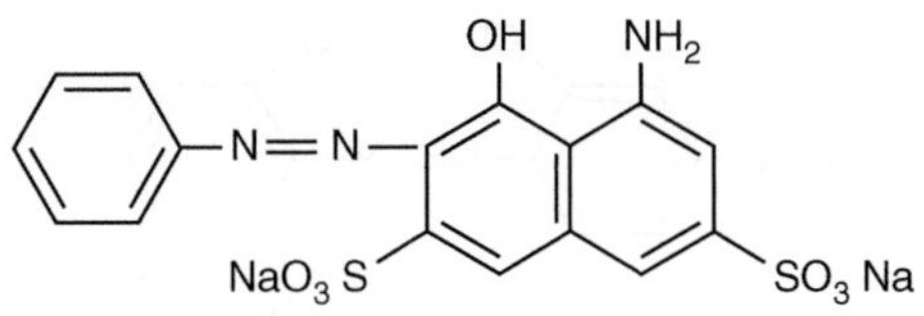

Figure 1.9   D&C Red 33.

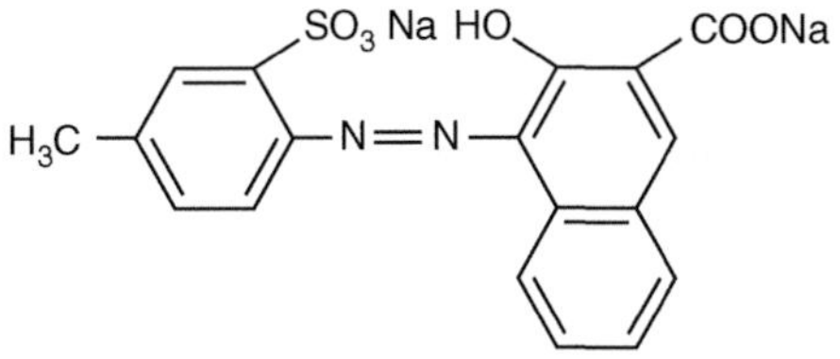

Figure 1.10   D&C Red 6.

dye contents, defined as the amount of dye in the lake, range from 12 to 40%. Figure 1.9 shows the chemical structure of a typical dye used to form a lake, in this case D&C Red 33. As the reader may see, the structure contains two sulfonic acid (SO_3) groups, which renders the material quite water-soluble.

Toners: Pigments produced by precipitating a water-soluble dye as a metal salt. Typical metals used for the precipitation include sodium, calcium, barium, and strontium. Toners are capable of complete insolubility without the aid of a substrate. In actual practice in the cosmetic industry, most toners are diluted with an inert substance to reduce their intensity, improve their dispersibility, or change their transparency. Materials commonly used for this purpose are barium sulfate, talc, and rosin. Figure 1.10 shows the structure of a toner. It contains both a sulfonic acid (SO_3) group and a carboxylic acid (COOH) group, which are hydrophilic and polar; therefore, as an acid dye, the toner is water-soluble.

True Pigments: Those colorants that are insoluble based on their chemical structure and constituent groups. They typically do not contain the normal substituent groups that promote water solubility, including sulfonic acid (SO_3) and carboxylic acid (COOH) groups. The true pigments are normally used without any diluents. A typical structure of a true pigment is shown in Figure 1.11.

All of the color additives described here find use in cosmetic products. Choosing the appropriate one for use in a given formulation is done based

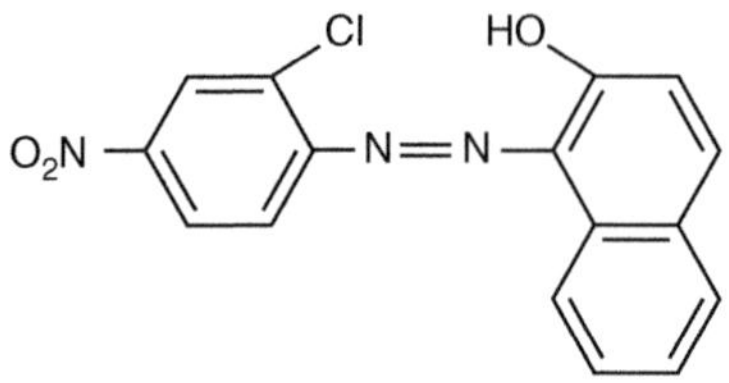

Figure 1.11   D&C Red 36.

Table 1.3   Organic pigment chemical classes.

Color additive	Chemical class
Carmine	Lake
FD&C Red No. 40 Al Lake	Lake
D&C Red No. 6	Toner
D&C Red No. 6 Ba Lake	Toner
D&C Red No. 7 Ca Lake	Toner
D&C Red No. 21 Al Lake	Lake
D&C Red No. 22 Al Lake	Lake
D&C Red No. 27 Al Lake	Lake
D&C Red No. 28 Al Lake	Lake
D&C Red No. 30 Al Lake	True pigment
D&C Red No. 33 Zr Lake	Lake
D&C Red No. 33 Al Lake	Lake
D&C Red No. 34 Ca Lake	Toner
D&C Red No. 36	True pigment
D&C Orange No. 5 Al Lake	Lake
FD&C Yellow No. 5 Al Lake	Lake
FD&C Yellow No. 6 Al Lake	Lake
FD&C Yellow No. 10 Al Lake	Lake
FD&C Blue No. 1 Al Lake	Lake
D&C Black No. 2	True pigment
D&C Black No. 3	True pigment
D&C Violet No. 2	True pigment

on several factors, or properties, which singularly or in combination offer various benefits and detriments depending on the intended usage. Due to the variations in terms of color and physical properties of the organic pigments, it is essential for the formulator to know which type each is before working with it. Table 1.3 identifies the commonly used organic pigments by chemical class.

Color Properties

As just stated, color additives exhibit properties that affect how they produce the desired color effect in cosmetics and toiletry products, which in turn affects their applicability. The following is a list of said properties:

Masstone: The ability of a color additive to impart a deep rich tone to a cosmetic product. This is particularly important in dark shades of lipstick and nail polish that contain no white.

Transparency: The property of a color additive that permits light and images to be transmitted from a substrate back through the color film.

Shade: As the name indicates, the ability of a color additive to impart a particular color to a finished product (e.g., blue shade red, green shade blue, red shade yellow, etc.).

Strength: The property of a color additive that provides intensity of color to a cosmetic product.

Bleed: The situation where a lake, toner, or true pigment has a very slight amount of solubility in the vehicle used to produce a particular cosmetic product. This slight solubility can potentially cause a staining of the skin or nail, or color flotation in the product itself.

Dispersibility: The measure of the ease with which a pigment can be incorporated into the base vehicle in a decorative cosmetic product (e.g. castor oil for lipstick, nitrocellulose for nail polish, talc for powders, water for liquid makeup, etc.).

Rheology: The measurement of a pigment's ability to affect the viscosity and flow properties of liquid cosmetic products. This is especially important in nail polish and liquid makeup.

Oil Absorption: Based on both their chemical structure and the type and amount of substrates and diluents present in lakes and toners, they will exhibit a wide range of oil absorptive powers. This property is significant, for example, in the formulation of lipsticks, as the oil absorption of a pigment has a major impact on the hardness of the stick.

Color Nomenclature

The identification of color additives can be very confusing, because different countries use different arbitrary systems to name them. Fortunately, however, there is a universally accepted system designed by the Society of Dyers and Colourists based in the United Kingdom, called the Colour Index. In this system, colorants are identified in two ways. First, a five-digit number is assigned to each separate color chemical moiety (toners may have a sixth digit added at the end following a colon to designate the precipitating metal salt). Second, each colorant is assigned a color name and number, corresponding to this five (or six)-digit Colour Index number.

For example, by US nomenclature, the most common organic toner used in decorative cosmetics is D&C Red 7, which is a calcium salt. This colorant is also used in industrial applications such as printing ink and plastics. In those applications, its designation as D&C Red 7 is meaningless, but users can easily recognize the colorant by its Colour Index system name and number, which are Pigment Red 57:1 and 15850:1, respectively. More specific nomenclature for color additives used in cosmetics and toiletries will be discussed in Chapter 2.

Organic vs. Inorganic

No discussion, particularly in today's marketplace, would be complete without a review of the difference between "organic" and "inorganic." While there is a lot of controversy in the food and cosmetics industries over the definition of "organic," it is very clear what it means in the world of chemistry. According to Merriam-Webster's New Collegiate Dictionary, "organic" is defined in two ways: primarily as "produced with or based on the use of fertilizer of plant or animal origin without the employment of chemically formulated fertilizers or pesticides," and secondarily as "relating to a branch of chemistry concerned with or containing Carbon compounds" (Merriam-Webster 2003). For the purposes of this book, the second definition will apply in all cases. Similarly, there are two dictionary definitions of "inorganic": "being composed of matter other than plant or animal" and "relating to a branch of chemistry concerned with substances not usually classed as organic" (Merriam-Webster 2003). Again, the second definition will apply for all purposes in this book.

Reference

Merriam-Webster (2003). *Webster's New Collegiate Dictionary*. Springfield, MA: The Merriam-Webster Company.

Chapter 2

Color Selection – Regulations

Editor's Note: The updates in Chapter 2 were kindly provided by James So, Regulatory Lead, Cosmetics of Sun Chemical Ltd.

This is one of the longer chapters in this book, and it is for good reason. The most complicated issue facing formulators on a daily basis is the minefield of intricate, ever-changing regulations pertaining to colorants found in seemingly myriad iterations throughout the world.

The color selection process begins with regulations. It is absolutely essential to ensure that any color additive selected for use in a decorative cosmetic or toiletry product meets the regulatory requirements of the countries where a product will be sold. Colorants are the most highly regulated chemicals used in cosmetic products. As the world has grown smaller – with the advent and perpetuation of technology, handheld and otherwise, bridging nearly all geographic distances and bringing people in disparate countries in ready contact with one another – it has become important for all cosmetic chemists, no matter where they are based, to have a thorough understanding of the regulations of the United States Food and Drug Administration (FDA) and the other major global markets: the European Union, Japan, and China.

In order to thoroughly address the subject, the material in this chapter will be divided into four sections:

1. The regulatory history in the United States.
2. The current regulatory landscape in the United States.

Coloring the Cosmetic World: Using Pigments in Decorative Cosmetic Formulations, Second Edition. Edwin B. Faulkner. Edited by Jane C. Hollenberg.
© 2021 John Wiley & Sons Ltd. Published 2021 by John Wiley & Sons Ltd.

3. The state of regulations in the European Union, Japan, and China.
4. Regulatory changes on the horizon in the United States, the European Union, Japan, and China.

The Regulatory History in the United States

George Washington is commonly known as the father of the United States. The first president, he was a war hero and shepherd of many of the notably "American" values that are still prized today. What isn't so commonly known is that Abraham Lincoln, America's 16th president, was the father of the FDA. In 1862, Lincoln authorized the establishment of the United States Department of Agriculture (USDA) and its subunit, the Bureau of Chemistry, which was the predecessor of the FDA (USDA n.d.).

The regulation of colors for use in food, drugs, and cosmetics began in the early part of the twentieth century. Because of the growing use of potentially dangerous color additives in food, such as copper acid orthoarsenite and lead chromate, the United States Congress enacted the Food and Drug Act of 1906, to be administered by the Bureau of Chemistry. The law established for the first time a list of approved color additives for use in food. These colorants were all "coal tar colors," the terminology used for synthetic, organic colors, based, at the time, on coal tar. The seven approved colorants are listed in Table 2.1 (Hochheiser 1982). Three are still approved for food use today: FD&C Red No. 3, FD&C Blue No. 2, and Ext. D&C Yellow No. 7, the first two for food only and the third for externally applied cosmetics.

In addition to establishing this positive list of approved colors, the Food and Drug Act also provided for the voluntary certification of individual batches of color additives, the first of which was certified on April 1, 1908. As an interesting historical note, it was the practice of the Bureau to test new food additives on people rather than animals. A group of volunteer men, affectionately called the "poison squad" (Figure 2.1), would eat these materials and then be observed in order to see what happened to them!

Between the enactment of the Act in 1906 and 1929, the year that saw the beginning of the Great Depression, another 10 colorants were added to the positive list of approved color additives. During the same period, the use of colorants in drug and cosmetic products became increasingly prevalent. In 1927, the Bureau of Chemistry was reorganized and renamed the Food, Drug and Insecticide Administration; in 1931, it was given its current name, the Food and Drug Administration. By 1937, the US government was certifying

Table 2.1   Original colors listed in the Food and Drug Act of 1906.

Original name	Current name
Amaranth	No longer permitted
Ponceau 3R	No longer permitted
Orange 1	No longer permitted
Erythrosine	FD&C Red No. 3
Naphthol Yellow S	Ext. D&C Yellow No. 7
Light green SF Yellowish	No longer permitted
Indigo Disulfo Acid sodium salt	FD&C Blue No. 2

Figure 2.1   The "poison squad."

230 metric tons of color additives annually. In 1940, the FDA was transferred from the USDA to the Federal Security Agency, which, in 1953, became the Department of Health, Education and Welfare, and is now the Department of Health and Human Services.

As a result of the dramatic increase in the number of end uses and the volume of colorants in food, drugs, and cosmetics, Congress passed the Food Drug and Cosmetic Act (FD&C Act) of 1938. This was a very comprehensive law, marking the first time that the US government had regulated the use of color additives in drugs and cosmetic products. It made the

Table 2.2   Definitions[a] of certifiable colors (FDA 2018a).

Food, Drug and Cosmetic Act of 1938

FD&C	Certifiable colors that are for use in coloring food, drugs, and cosmetics.
D&C	Certifiable colors that are for use in coloring drugs and cosmetics, including those in contact with mucous membranes and those that are ingested.
Ext. D&C	Certifiable colors that are for use in coloring drugs and cosmetics that do not come in contact with mucous membranes and those that are not ingested.

[a]Original definitions – no longer applicable in some cases.

certification of coal tar colors mandatory and created the three categories of color additives that are still in use today: FD&C Colors, D&C Colors, and Ext. D&C Colors (Table 2.2), but these categories may no longer apply in some cases. For example, some FD&C colorants are no longer permitted for use in cosmetic and toiletry products, such as FD&C Red No. 3 and FD&C Blue No. 2. The 1938 law also specified which color additives were permitted in each of the product categories of food, drugs, and cosmetics and provided well-defined chemical identities and chemical specifications for each color on the approved list. Finally, it set up regulations with respect to the manufacturing, labeling, certifying, and selling of color additives (FDA 2018b).

By the 1950s, although many new colorants had been added to the list of approved colorants for foods, drugs, and cosmetics, great concern developed about the safety of these additives. This concern was twofold. First, there were a number of instances where people became ill after ingesting foods that contained "high levels" of color additives. Second, the results of phar-macological tests performed by the FDA showed potential problems with some colorants, such as D&C Red No. 9 and D&C Red No. 19.

Disagreements developed between the industry and the FDA over safe use levels and the definition of the term "harmless." The FDA held to the view that a color additive must be safe, irrespective of its use level. This became the definition of the term "harmless per se." The industry, however, took the position that a color additive should be safe when used at the level necessary to produce the desired color effect.

All of this controversy resulted in a compromise that brought changes to the Food and Drug Act of 1938, which came to be called the Color Additive Amendments of 1960. These amendments allowed the continued use of colorants on the existing positive list of approved color additives, pending the completion of safety testing. The responsibility, including all associated costs, for this testing was shifted from the FDA to industry. The 1960 amendments also authorized the FDA to set use limits on the amount of colorant in finished products, thereby eliminating the controversy of the term "harmless per se." In addition, the FDA was allowed to decide which color additives had to continue to be certified and which could be exempt from certification.

The provision regarding safety testing of the color additives in use at the time called for industry to present scientific data demonstrating the safety of each colorant in order to have it permanently listed for use. Until these data were presented, a color additive was considered provisionally listed and could continue to be marketed. The FDA was given the authority to extend the provisional listing of each colorant until the safety testing was completed and the data were submitted.

Because of the time and financial resources required to complete the testing of a color additive, many colorants were removed from the approved list by default, as industry was unwilling to spend the money required to test those having small usage volumes. Many other color additives have been delisted in the decades since 1960 because of unfavorable results from safety testing. In 1959, there were 116 synthetic organic colorants in use for cosmetics. As a result of the requirements of the 1960 amendments, only 36 are permitted for this use today. Meanwhile, only three certifiable colorants have been added to the positive list in the past 60 years: FD&C Red No. 40, D&C Black No. 2, and D&C Black No. 3.

Users should appreciate the fact that the major provisions of the color regulations currently in force today are a result of all this activity over more than a century.

The Current Regulatory Landscape in the United States

There are two important FDA definitions that must be understood before proceeding further: **cosmetic** and **color additive** (FDA 2018a). Both are derived from the FD&C Act, Section 201:

> The term **cosmetic** means (1) articles intended to be rubbed, poured, sprinkled, or sprayed on, introduced into, or otherwise applied to the human body or any part thereof for cleansing, beautifying, promoting attractiveness, or altering the appearance, and (2) articles intended for use as a component of any such articles; except that such term shall not include soap.

A **color additive** is any material, not exempted under section 201 (t) of the Act, that is a dye, pigment, or other substance made by a process of synthesis or similar artifice, or extracted, isolated, or otherwise derived, with or without intermediate or final change of identity, from a vegetable, animal, mineral, or other source and that, when added or applied to a food, drug, or cosmetic or to the human body or any part thereof is capable (alone or through reaction with another substance) of imparting a color thereto. Hair dyes were exempted.

Now, on to the specifics of today's color regulations.

The permanent list of approved color additives is divided into two parts:

Exempt Colors: The exempt color additives are defined chemically as the nonsynthetic organic and the inorganic colors. Batch certification of the exempt colorants is not required. However, there are purity specifications that must be followed by both the manufacturing and the consuming companies. The specifications are different for each color additive on the Exempt list, as listed in the Code of Federal Regulations (FDA 2018c). (Exempt colors are taken up again later in this chapter.)

Certifiable Colors: The certifiable colors are defined chemically as synthetic organic colorants. These color additives must be batch-certified, meaning that the manufacturer of the colorant must submit a sample of each batch to the FDA for analysis. If it meets the published purity specifications, the FDA issues a certification number for the batch to the manufacturing company. The manufacturing company must display this certification number on the package in which the colorant is shipped and on all of the associated paperwork. In addition, the food, drug, or cosmetic company that purchases the batch must keep a record of where it was used in the company's production.

The certifiable colors are further divided into two categories.

Primary Colors: The primary colors are those colors that are pure. That is, they may not contain any diluents. They can be dyes or pigments depending on their individual chemistry.

Lakes: The lakes are those primary dyes that are absorbed on to one or more of the approved substrates or those primary toners/true pigments that are reduced (extended) with one or more of the approved substrates A list of the approved substrates may be found in Table 2.3. All of the lakes are pigments.

Table 2.3   Permitted substrates for cosmetic lakes (FDA 2018a).

1. Barium sulfate
2. Gloss white
3. Talc
4. Zinc oxide
5. Titanium dioxide
6. Clay
7. Aluminum hydrate
8. Aluminum benzoate
9. Calcium carbonate
10. Rosin

As a result of the 1938 Act and the 1960 Amendments, the current nomenclature conventions are as follows:

Certifiable Primary Colors: Each of the colorants in this category has both a color name and a number assigned to it. If it is approved for all three categories of use, the name is preceded by the letters "FD&C." If it is approved for use only in drugs and cosmetics, the name is preceded by the letters "D&C." If approved for use in externally applied drugs and cosmetics, the name is preceded by the letters "Ext. D&C." Examples are FD&C Yellow No. 5, D&C Red No. 6, and Ext. D&C Violet No. 2. As previously mentioned, however, there are some anomalies in this naming system. Due to the safety testing results presented for some of the colorants, their names are not always specifically accurate. For example, FD&C Red No. 3 is no longer permitted for use in cosmetic or toiletry products, even though by classification name it should be. Likewise, D&C Red No. 34 is now an "external only" color. One must always consult the specific paragraph of the current US Code of Federal Regulations Title 21, Parts 1–99, before using a color additive in a cosmetic or toiletry product.

Certifiable Color Lakes: The nomenclature for lakes follows the same conventions and cautions that apply to the primary colors, with the additional rule that the lake of the color must contain the name of the precipitating metal and the word "Lake." An example would be FD&C Yellow No. 5 Al Lake.

Exempt Colors: The names assigned to the exempt colors do not follow any organized name or numbering system. They simply describe the color based on its common name (Ultramarines, Carmine) or chemical structure (Titanium Dioxide, Iron Oxides).

As another result of the 1938 Act and 1960 Amendments, there are several restrictions placed on the use of both certifiable and exempt colors:

Use Restrictions: These dictate in which type of cosmetic or toiletry product a particular color additive may be used. The categories are: (i) Use in any cosmetic product; and (ii) Use in externally applied cosmetic products. Externally applied cosmetics are those that do not come in contact with a mucous membrane. (iii) Eye area use is only permitted in either category if specifically mentioned.

Quantity Restrictions: These dictate how much of a specific color additive may be used in a cosmetic or toiletry product. In the absence of a specific amount, the regulations state that a color additive "shall be used in an amount consistent with good manufacturing practice."

A listing of the certifiable colors currently approved for cosmetics and toiletries, along with their associated use and quantity restrictions, can be found in Table 2.4, while a listing of the approved exempt colors may be found in Table 2.5.

The amount of color certified has grown significantly over the years, from 230 metric tons in 1937 to more than 6000 in fiscal year 1998, based on FDA-published statistics. A listing of the 2018 individual colorant certification data can be found in Table 2.6.

The certification data for each color additive is a summary for all uses with no breakdown by end use application. The vast majority of FD&C primary and lake colorants are used in the food industry, not in cosmetics and toiletries. As a reinforcement that the number of colorants actually used in cosmetics is small, it can be seen that the certification data for the D&C primary and lake colorants are centered on a few of them.

Because of the long history of regulatory changes in the United States over the last 110 years, there are several confusing aspects in the application of the regulations that need to be brought to the reader's attention before moving on to international regulation of colors:

As noted earlier, but worth stating again, not all FD&C Colors are permitted for use in cosmetics.

Also noted earlier, not all D&C and exempt colorants are permitted for use in mucous membrane and ingested cosmetics.

Only synthetic iron oxides are permitted for use in cosmetics; natural ones are not permitted.

Table 2.4   Permitted certifiable colors (FDA 2018d).

Name	Use restrictions	Quantity restrictions
FD&C Green No. 3	Not for use in the area of the eye	Consistent with GMP[a]
FD&C Yellow No. 5	None	Consistent with GMP
FD&C Yellow No. 6	Not for use in the area of the eye	Consistent with GMP
FD&C Red No. 4	External only	Consistent with GMP
FD&C Red No. 40	None	Consistent with GMP
FD&C Blue No. 1	None	Consistent with GMP
D&C Green No. 5	None	Consistent with GMP
D&C Green No. 6	External only	Consistent with GMP
D&C Green No. 8	External only	0.01% by weight of finished product[b]
D&C Yellow No. 7	External only	Consistent with GMP
D&C Yellow No. 8	External only	Consistent with GMP
D&C Yellow No. 10	Not for use in the area of the eye	Consistent with GMP
D&C Yellow No. 11	External only	Consistent with GMP
D&C Red No. 6	Not for use in the area of the eye	Consistent with GMP
D&C Red No. 7	Not for use in the area of the eye	Consistent with GMP
D&C Red No. 17	External only	Consistent with GMP
D&C Red No. 21	Not for use in the area of the eye	Consistent with GMP
D&C Red No. 22	Not for use in the area of the eye	Consistent with GMP
D&C Red No. 27	Not for use in the area of the eye	Consistent with GMP
D&C Red No. 28	Not for use in the area of the eye	Consistent with GMP
D&C Red No. 30	Not for use in the area of the eye	Consistent with GMP
D&C Red No. 31	External only	Consistent with GMP
D&C Red No. 33	Not for use in the area of the eye	3% in lipstick[b]
D&C Red No. 34	External only	Consistent with GMP
D&C Red No. 36	Not for use in the area of the eye	3% in lipstick[b]
D&C Orange No. 4	External only	Consistent with GMP
D&C Orange No. 5	Not for use in the area of the eye	5% in lipstick[b]
D&C Orange No. 10	External only	Consistent with GMP
D&C Orange No. 11	External only	Consistent with GMP
D&C Blue No. 4	External only	Consistent with GMP
D&C Brown No. 1	External only	Consistent with GMP
D&C Violet No. 2	External only	Consistent with GMP
Ext. D&C Violet No. 2	External only	Consistent with GMP
Ext. D&C Yellow No. 7	External only	Consistent with GMP
D&C Black No. 2	May be used in eyeliner, brush-on brow, eyeshadow, mascara, lipstick, blushers, rouge, makeup and foundation, and nail polish	Consistent with GMP
D&C Black No. 3	May be used in eyeliner, eyeshadow, mascara, and face powder	Consistent with GMP

[a]GMP, good manufacturing practice.
[b]Based on 100% dye.

Table 2.5   Use restrictions on exempt colors (FDA 2018c).

Name	Ingested use	External use	Eye use	Comments
Aluminum Powder	No	Yes	Yes	
Annatto	Yes	Yes	Yes	
Bismuth Citrate	No	Yes	Yes	2% in hair color
Bismuth Oxychloride	Yes	Yes	Yes	
Bronze Powder	Yes	Yes	Yes	
Caramel	Yes	Yes	Yes	
Carmine	Yes	Yes	Yes	
Carotene	Yes	Yes	Yes	
Chlorophyllin Cu Complex	Yes	No	No	For use only in dentifrices at 0.1% max
Chromium Hydroxide Green	No	Yes	Yes	
Chromium Oxide Green	No	Yes	Yes	
Copper Powder	Yes	Yes	Yes	
Dihydroxyacetone	No	Yes	No	
Disodium EDTA	No	Yes	No	
Ferric Ammonium Ferrocyanide	No	Yes	Yes	
Ferric Ferrocyanide	No	Yes	Yes	
Guanine	Yes	Yes	Yes	
Guaiazulene	No	Yes	No	
Henna	No	Yes	No	For use in hair coloring only
Iron Oxides	Yes	Yes	Yes	
Lead Acetate	No	[a]	No	Status pending
Manganese Violet	Yes	Yes	Yes	
Mica	Yes	Yes	Yes	Max 150 microns
Pyrophyllite	No	Yes	Yes	
Titanium Dioxide	Yes	Yes	Yes	
Ultramarines	No	Yes	Yes	
Zinc Oxide	Yes	Yes	Yes	
Silver	No	Yes	No	For nail polish only at 1% max
Luminescent Zinc Sulfide	No	Yes	No	Only occasional use, except for nail polish

[a]Delisted October 2018; listing reinstated pending review of objections to delisting April 1, 2019.

There are no FDA specifications for bacteria or mold, but the cosmetic industry sets its own specification at 100 colonies per gram with no pathogens present.

If a cosmetic product carries a drug claim, for example a lipstick with a sun protection factor (SPF), the color additives in the product must comply

Table 2.6 US FDA certification data FY 2018 (FDA 2018f).[a]

Name	Amt. certified	Name	Amt. certified
FD&C primary colors		FD&C lakes	
FD&C Green No. 3	23 053.51	FD&C Green No. 3	0
FD&C Yellow No. 5	4 004 682.11	FD&C Yellow No. 5	1 298 052.15
FD&C Yellow No. 6	4 801 564.52	FD&C Yellow No. 6	1 294 612.91
FD&C Red No. 3	261 335.13	FD&C Red No. 40	2 426 341.01
FD&C Red No. 4	31 044.45	FD&C Blue No. 1	524 174.13
FD&C Red No. 40	5 392 824.33	FD&C Blue No. 2	186 504.40
FD&C Blue No. 1	962 334.18		
FD&C Blue No. 2	370 736.10		
D&C Primary Colors		D&C lakes	
D&C Black No. 2	107 376.71	D&C Yellow No. 5	2998.26
D&C Black No. 3	0	D&C Yellow No. 6	4673.75
D&C Green No.5	4953.40	D&C Yellow No. 10	85 656.49
D&C Green No.6	8738.10	D&C Red No. 6	353 101.60
D&C Green No.8	19 295.18	D&C Red No. 7	717 237.68
D&C Yellow No. 8	8106.06	D&C Red No. 21	60 784.96
D&C Yellow No. 10	61 967.24	D&C Red No. 22	53 532.02
D&C Yellow No. 11	13 601.72	D&C Red No. 27	303 476.27
D&C Red No. 6	56 891.94	D&C Red No. 28	268 254.11
D&C Red No. 7	86 202.77	D&C Red No. 30	65 620.65
D&C Red No. 17	6806.61	D&C Red No. 33	41 211.99
D&C Red No. 21	1185.84	D&C Red No. 34	62 906.78
D&C Red No. 22	7261.83	D&C Orange No. 5	1301.50
D&C Red No. 27	5235.22		
D&C Red No. 28	21 888.85		
D&C Red No. 30	27 660.03		
D&C Red No. 31	0		
D&C Red No. 33	82 019.34		
D&C Red No. 34	8868.63		
D&C Red No. 36	9265.41		
D&C Orange No. 4	23 296.73		
D&C Orange No. 5	3741.83		
D&C Violet No. 2	8706.60		

[a]Certification for all uses: foods, drugs, cosmetics, and medical devices.

with the FDA drug regulations, not the cosmetic ones. This means that the formulator must go to the drug paragraph of the color regulations in the Code of Federal Regulations Title 21, Parts 73 and 74 in order to determine how to use a particular color additive in the product.

The State of Regulations in the European Union, Japan, and China

The cosmetic industry is a truly global business. Cosmetic and toiletry products made in one country will likely be sold in numerous others. Businesses within the industry are always attempting to penetrate new markets, to widen their catalogue of offerings, and to acquire new dynamic capabilities that will broaden and strengthen their produce portfolios. The color additives used in these products must meet the regulatory requirements of all the countries in which the cosmetics will be sold, however many that may be. Unfortunately, there are almost as many sets of color additive regulations as there are countries in the world. On the other hand, the color regulations of most countries will follow those of the four major world cosmetic markets: the United States, the European Union, Japan, and China. The first formal Chinese cosmetics regulations, The Hygienic Standard of Cosmetics, were promulgated in 2007; the portion pertaining to color additives is essentially identical to the EU regulations. This Standard has since been replaced by the Safety and Technical Standards for Cosmetics (2015 version), which came into force in December 2016. In order to simplify the material presented here, the similarities and differences amongst the color additive regulations in these four markets will be explored. For further details on the specific regulations for other countries, the reader is referred to *The International Color Handbook*, 4th edition, published by the Personal Care Products Council, Washington, DC (Rosholt 2007).

Unfortunately, the similarities amongst the four markets' color additive regulations are outnumbered by the differences. The areas of similarity are:

Positive List: All four markets have positive lists for the color additives that are permitted for use in cosmetic and toiletry products.

Primary Colors and Lakes: All four markets make distinctions between primary colors and lakes. The regulations provide details on which lakes are permitted.

Exempt Color Additives: All of the commonly used FDA exempt color additives are on the positive lists in the European Union/China (here melded into a single group due to their similarities) and Japan. Table 2.7 lists the common US exempt colorants and shows their corresponding names in the other two markets. It must be noted that some of the more

Table 2.7   Common US exempt color additives and their EU/Chinese and Japanese counterpart names.

United States (FDA 2018c)	European Union/China (European Commission 2009/NMPA 2015)	Japan (MHLW 2003)[a]
Aluminum Powder	77000	Aluminum Powder
Carmine	75470	Carmine
Chromium Hydroxide Green	77289	Hydrated Chromium Oxide
Chromium Oxide Green	77288	Chromium Oxide Green
Ferric Ammonium Ferrocyanide	77510	Ferric Ammonium Ferrocyanide
Iron Oxides	77491	Red Oxide of Iron
Iron Oxides	77492	Yellow Oxide of Iron
Iron Oxides	77499	Black Oxide of Iron
Manganese Violet	77742	Manganese Violet
Titanium Dioxide	77891	Titanium Dioxide
Ultramarines	77007	Ultramarine
Zinc Oxide	77947	Zinc Oxide

[a]These are the common reference names. Due to the major revision of the Japanese color regulations in 2004, these colorants are no longer officially regulated there.

obscure exempt colorants may not be on the positive list in all four markets.

Use Restrictions: All four markets restrict the use of some colors in certain types of cosmetic and toiletry products but these are not uniform in all jurisdictions. For example, ferric ammonium ferrocyanide is limited to external use only in the United States, but no such restriction exists in the European Union/China.

Specifications: All four markets have chemical specifications for the color additives on their positive lists that may differ among them. An example is that D& C Yellow No. 10 and CI 47005 are both Quinoline Yellow, but the former is specified in the United States as composed of not less than 75% of the monosodium salt and the latter in the European Union as not less than 80% of the disodium salt.

Precipitants: All four markets specify what precipitants are allowed for the manufacture of lakes. The United States has a general positive list, allowing any of the precipitants to be used with any of the primary colors. Although Japan does not have a general positive list, in the Japanese regulations, precipitants are specified for each of the colorants individually in the color additive positive list, divided into two categories. The first contains the precipitants that are specified in the chemical

Table 2.8   Approved cosmetic lake precipitants.

United States (FDA 2018e)	European Union/China (European Commission 2009/NMPA 2015)	Japan (MHLW 2003)[a]
Sodium	Sodium	Sodium
Potassium	Potassium	Potassium
Aluminum	Aluminum	Aluminum
Barium	Barium[b]	Barium
Calcium	Calcium	Calcium
Strontium	Strontium[b]	Strontium
Zirconium	Zirconium[b]	Zirconium

[a]Not all precipitants are permitted with all color additives.
[b]Not permitted with certain color additives.

definition of each colorant. The second contains the precipitants that are permitted for the manufacture of lakes. Again, this list is color additive specific. In the European Union/China, the innocuous metals approved in the United States and Japan are generally accepted, but some of these precipitants – zirconium, barium, and strontium – are restricted for use with certain colors. Table 2.8 provides a comparison among the regions of the approved precipitants for lakes.

The differences amongst the regulations in the four markets make development of cosmetic formulations very challenging. The major ones are:

Nomenclature: As described previously, the US system is an arbitrary one that does not relate to any other system. The Japanese system, likewise, is arbitrary. It provides for the use of a descriptive color name followed by a number (e.g., Red 201). The European Union/China system uses Color Index numbers to identify approved color additives. Examples of all these systems with regard to certifiable colorants can be found in Table 2.9.

Approved Color Additives: Even though all four markets have positive lists, the colorants that are permitted in each vary considerably. The positive lists in the European Union/China and Japan each contain far more colorants than the FDA positive list, so formulators have more choice in creating a particular color shade if their product is to be sold in one of these markets.

Listing Status: The United States and the European Union/China have both provisional and permanent lists of approved color additives. As the names imply, the permanently listed color additives are those for which the regulatory bodies are satisfied that the colorants are safe for use in cosmetic

Table 2.9 Common US certifiable color additives and their EU/Chinese and Japanese counterpart names.

United States (FDA 2018d)	European Union/China (European Commission 2009/NMPA 2015)	Japan (MHLW 2003)[a]
FD&C Red No. 40 Al Lake	16035	Not permitted
D&C Red No. 6 Ba Lake	15850	Not permitted
D&C Red No. 7 Ca Lake	15850	Red 202 Mixture
D&C Red No. 21 Al Lake	45380	Not permitted
D&C Red No. 22 Al Lake	45380	Red 230 (1) Lake
D&C Red No. 27 Al Lake	45410	Not permitted
D&C Red No. 28 Al Lake	45410	Red 104 (1) Lake
D&C Red No. 30 Al Lake	73360	Red 226 Mixture
D&C Red No. 30	73360	Red 226
D&C Red No. 31 Ca Lake	15800	Red 219
D&C Red No. 33 Al Lake	17200	Red 227 Lake
D&C Red No. 33 Zr Lake	Not permitted	Not permitted
D&C Red No. 34 Ca Lake	15880	Red 220 Mixture
D&C Red 36	12085	Red 228
D&C Orange No. 5 Al Lake	45370	Not permitted
FD&C Yellow No. 5 Al Lake	19140	Yellow 4 Lake
FD&C Yellow No. 6 Al Lake	15985	Yellow 5 Lake
D&C Yellow No. 10 Al Lake, primarily a monosulfonate	47005, a disulfonate; differs from D&C Yellow No. 10	Yellow 203 Lake
FD&C Blue No. 1 Al Lake	42090	Blue 1 Lake
D&C Black No. 2	77266	Carbon Black

[a]Aluminum lakes of some color additives are not permitted.

and toiletry products. The provisionally listed color additives are those for which some safety studies remain pending or the results of such tests are under review. The Japanese regulations have only a permanent list of color additives.

Lake Definition: The EU/Chinese and Japanese regulations use the classical chemical definition of a lake as described earlier in this chapter. That is, a lake is a pigment that is produced by absorbing a water-soluble dye on to an insoluble, inorganic substrate. The US definition differs considerably; while the US regulations embrace the classical definition, they also define a lake as any color additive, dye, or pigment that is extended (reduced) with one of the 10 approved substrates. This difference leads to a considerable amount of confusion. Some important colorants, particularly the reduced toners, which are considered lakes under the US regulations, would appear to not be permitted as lakes under the Japanese ones. But, since they are not true chemical lakes, they are permitted for use in Japan. Two major examples of this anomaly are D&C Red No. 7 Ca Lake and D&C Red 30 Al Lake.

Specifications: Specifications, like the colorants on the positive lists, are not the same for all four markets. The major differences are found in the chemical purity and allowable level of contaminants.

Certification: Only the United States has a mechanism for testing and subsequent certification of each batch of synthetic organic colors for cosmetics. Japan does have a batch certification program for food colors. The US certification process is administered by the FDA. All cosmetic products destined for sale in the United States must use certified color additives.

Substrates: The United States is the only market that has a positive list for substrates. In the European Union/China and Japan, any chemical that is used as a substrate in the manufacture of a cosmetic color must be acceptable as a cosmetic ingredient in that market. The European Union does not have a positive list for all cosmetic ingredients, but it does have prohibited and restricted lists, like China and Japan. Table 2.10 lists the substrates that are permitted in the United States along with their status in the European Union/China and Japan.

Use Restrictions: While all four markets have use restrictions, these restrictions vary amongst them. The most notable difference is that the US regulations permit only some synthetic organic colorants to be used in eye makeup, while many more of these types of colorants are permitted for this use in both the European Union/China and Japan. Table 2.11 lists the most commonly used US color additives along with their corresponding use restrictions in all four markets.

Quantity Restrictions: The United States and the European Union/China each have restrictions on the quantity of certain colors permitted for use in cosmetic products, but these restrictions differ between the two markets. The United States does not list specific quantity limits on all colors. As mentioned earlier, for those that do not have a specific amount listed, the regulations will state that the color "shall be used in amounts consistent with good manufacturing practice (GMP)." Japan does not have any quantity restrictions. Similarly, the EU/Chinese regulations do not list specific use amounts for every color. But, unlike with the United States, there is not a similar GMP quantity use clause. And while Japan, again, does not have quantity restrictions per se, it must be noted that there are commonly accepted use levels in the Japanese market. Table 2.12 lists the quantity restrictions placed on some commonly used US color additives.

Table 2.10 Substrates approved in the United States and their EU/Chinese and Japanese statuses.

United States (FDA 2018e)	European Union/China (European Commission 2009/NMPA 2015)	Japan (MHLW 2003)
Barium sulfate	No restrictions	No restrictions
Gloss white	No restrictions	No restrictions
Talc	No restrictions	No restrictions
Zinc oxide	No restrictions	No restrictions
Titanium dioxide	No restrictions	No restrictions
Clay	No restrictions	No restrictions
Aluminum hydrate	No restrictions	No restrictions
Aluminum benzoate	No restrictions	No restrictions
Calcium carbonate	No restrictions	No restrictions
Rosin	No restrictions	No restrictions

To demonstrate how the differences in the global color regulations can make formulating cosmetic products a challenge, look back at Table 2.9. This shows the most common certifiable colorants used in the US market, along with their regulatory status in the European Union/China and Japan. If a colorant is approved for use in the European Union or Japan, its name in that market is shown. As can be seen, several important colors, the most notable of which is D&C Red No. 6 Ba Lake – the major yellow shade certifiable red used in US formulations – are not permitted for use in Japan.

Fortunately for the sanity of the cosmetic chemist, there are some other colorants on the US list that are acceptable in Japan. To replace D&C Red No. 6 Ba Lake, the formulator can use the primary color, D&C Red No. 6, in its place. Although some of these colors have not been widely used in the past, their use has become more prevalent as cosmetic companies seek to develop universally acceptable formulations. Table 2.13 includes these colorants in a more complete list of color additives that are permitted in all four markets.

As a final note on regulations, the current official publications of the US FDA, the European Union, China, and Japan must be consulted before using any color additives in a cosmetic or toiletry product. While the regulatory situation portrayed here is accurate at the time of writing, regulations change on a regular basis. So, the information presented here may have been revised by the various regulatory bodies in the intervening time between this writing and your reading.

Table 2.11 US certified color additives: cosmetic use restrictions.

Color additive	United States (FDA 2018d)	European Union/China (European Commission 2009/NMPA 2015)	Japan (MHLW 2003)
FD&C Red No. 40 Al Lake	No restrictions	No restrictions	Not permitted
D&C Red No. 6 Ba Lake	Not for use in the area of the eye	No restrictions	Not permitted
D&C Red No. 7 Ca Lake	Not for use in the area of the eye	No restrictions	Red 202 as a mixture
D&C Red No. 21 Al Lake	Not for use in the area of the eye	No restrictions	Not permitted
D&C Red No. 22 Al Lake	Not for use in the area of the eye	No restrictions	Red 230 (1) Lake
D&C Red No. 27 Al Lake	Not for use in the area of the eye	No restrictions	Not permitted
D&C Red No. 28 Al Lake	Not for use in the area of the eye	No restrictions	Red 104 (1) Lake
D&C Red No. 30 Al Lake	Not for use in the area of the eye	No restrictions	Red 226 as a mixture
D&C Red No. 30	Not for use in the area of the eye	No restrictions	No restrictions
D&C Red No. 31 Ca Lake	External only	External only	Nail and hair use only
D&C Red No. 33 Al Lake	Not for use in the area of the eye	No restrictions	No restrictions
D&C Red No. 33 Zr Lake	Not for use in the area of the eye	Not permitted	Not permitted
D&C Red No. 34 Ca Lake	External only	No restrictions	No restrictions
D&C Red 36	Not for use in the area of the eye	No restrictions	No restrictions
D&C Orange No. 5 Al Lake	Not for use in the area of the eye	No restrictions	Not permitted
FD&C Yellow No. 5 Al Lake	No restrictions	No restrictions	No restrictions
FD&C Yellow No. 6 Al Lake	Not for use in the area of the eye	No restrictions	No restrictions
D&C Yellow No. 10 Al Lake	Not for use in the area of the eye	No restrictions as CI 47005	No restrictions
FD&C Blue No. 1 Al Lake	No restrictions	No restrictions	No restrictions
D&C Black No. 2	Permitted uses listed; most color cosmetics	No restrictions	No restrictions; under industry standard

Table 2.12 US certified color additives: cosmetic quantity restrictions.

Color additive	United States (FDA 2018d)	European Union/China (European Commission 2009/NMPA 2015)	Japan (MHLW 2003)
FD&C Red No. 40 Al Lake	Consistent with GMP amounts	No restrictions	Not permitted
D&C Red No. 6 Ba Lake	Consistent with GMP amounts	No restrictions	Not permitted
D&C Red No. 7 Ca Lake	Consistent with GMP amounts	No restrictions	N/A[a]
D&C Red No. 21 Al Lake	Consistent with GMP amounts	No restrictions	Not permitted
D&C Red No. 22 Al Lake	Consistent with GMP amounts	No restrictions	N/A
D&C Red No. 27 Al Lake	Consistent with GMP amounts	No restrictions	Not permitted
D&C Red No. 28 Al Lake	Consistent with GMP amounts	No restrictions	N/A
D&C Red No. 30 Al Lake	Consistent with GMP amounts	No restrictions	N/A
D&C Red No. 30	Consistent with GMP amounts	No restrictions	N/A
D&C Red No. 31 Ca Lake	Consistent with GMP amounts	No restrictions	N/A
D&C Red No. 33 Al Lake	Maximum 3% in lipsticks	No restrictions	N/A
D&C Red No. 33 Zr Lake	Maximum 3% in lipsticks	Not permitted	Not permitted
D&C Red No. 34 Ca Lake	Consistent with GMP amounts	No restrictions	N/A
D&C Red 36	Maximum 3% in lipsticks	Maximum 3% in finished product	N/A
D&C Orange No. 5 Al Lake	Maximum 5% in lipsticks	No restrictions	Not permitted
FD&C Yellow No. 5 Al Lake	Consistent with GMP amounts	No restrictions	N/A
FD&C Yellow No. 6 Al Lake	Consistent with GMP amounts	No restrictions	N/A
D&C Yellow No. 10 Al Lake	Consistent with GMP amounts	No restrictions as CI 47005	N/A
FD&C Blue No. 1 Al Lake	Consistent with GMP amounts	No restrictions	N/A
D&C Black No. 2	Consistent with GMP amounts	No restrictions	N/A

[a] Not applicable, as Japan does not have quantity limitations.

Table 2.13   Universally acceptable color additives.

United States (FDA 2018a)	European Union/China (European Commission 2009/NMPA 2015)	Japan (MHLW 2003)
D&C Red No. 6	15850	Red 201
D&C Red No. 7 Ca Lake	15850	Red 202 (mixture)
D&C Red No. 22 Al Lake	45380	Red 230 (1) Lake
D&C Red No. 28 Al Lake	45410	Red 104 (1) Lake
D&C Red No. 30 Al Lake	73360	Red 226 (mixture)
D&C Red No. 30	73360	Red 226
D&C Red No. 31 Ca Lake	15800	Red 219
D&C Red No. 33 Al Lake	17200	Red 227 Lake
D&C Red No. 34 Ca Lake	15880	Red 220 Mixture
D&C Red 36	12085	Red 228
FD&C Yellow No. 5 Al Lake	19140	Yellow 4 Lake
FD&C Yellow No. 6 Al Lake	15985	Yellow 5 Lake
FD&C Blue No. 1 Al Lake	42090	Blue 1 Lake
D&C Black No. 2	77266	Carbon Black

Regulatory Changes in China, the United States, and the European Union

As was noted at the beginning of this chapter, regulations are ever-changing. This final section is a brief review of the changes that have occurred since the last edition of this book was published in 2012 and of regulatory issues evolving at the time of writing. There have been no changes to the color regulations in Japan, and none are anticipated.

China

As mentioned earlier, The Hygienic Standard of Cosmetics 2007 has been replaced by the Safety and Technical Standards for Cosmetics (2015 version). The new Standard came into force in December 2016. Again, from the color perspective, Table 2.6 contains a list of colorants approved in cosmetic products, along with substance identification, application areas, and other restrictions and requirements. One particular point to note is that the purity requirements for some colorants are extracted from either the EU or FDA specifications, or even from both sets for several colorants. Furthermore, a new Inventory of Existing Cosmetic Ingredients in China (IECIC) was issued by the China Food and Drug Administration (CFDA) in 2015, and was updated again in 2021. This is a list of existing cosmetic ingredients that have already been used in cosmetic products in China. Any substances not listed in the IECIC are regulated as new cosmetic ingredients that must be approved by the CFDA prior to being used in cosmetic products in China.

Please note that in 2018, the CFDA was merged into the State Administration for Market Surveillance (SAMR), and its name was changed to the National Medical Products Administration (NMPA).

United States

There is a potential change involving lakes in the United States. All of the lakes, with the exception of FD&C Red No. 40, are still provisionally listed. In the mid-1990s, the FDA published its draft regulations for lakes, which contained one revision that would have a dramatic impact on the color points and economics of the D&C lakes, but would not affect the FD&C ones. This provision would require that all D&C lakes be made from previously certified straight color. Currently, most of the D&C lakes can be made in situ, which means that the dye can be made and then precipitated as a pigment all in one step. If the change were to be made, the straight color would have to be made and certified, then returned to the production vessel, redissolved, and laked. This would cause the color point to change slightly and double the manufacturing cost for lakes. After receiving numerous objections from color additive and cosmetic manufacturers, the FDA fell silent on the draft regulations, and they have never been subsequently promulgated. However, this situation is like a dormant volcano; one doesn't expect it to erupt anytime soon, but the potential is always there.

Another regulatory action affecting a cosmetic pigment is the State of California's classification of Titanium Dioxide as an inhalation carcinogen under Proposition 65 (OEHHA 2011). Details follow the discussion of EU activities regarding Titanium Dioxide.

European Union

In 2009, the European Union developed a new regulatory document titled *Regulation (EC) No. 1223/2009 of the European Parliament and of the Council of 30 November 2009 on Cosmetic Products* (European Commission 2009). The regulation came into force in July 2013. From the color perspective, Annex IV contains a list of colorants allowed in cosmetic products. The list contains substance identification including the Color Index number, as well as the conditions for use (e.g., the product type and maximum concentration in ready-for-use preparations, if applicable.) There are also purity criteria inserted for some colorants. For those colorants which were already approved as food colors, the purity criteria are the same as set out in Commission Directive 95/45/EC, which lays down specific purity criteria concerning colors for use in foodstuffs (European Commission 2009).

Titanium Dioxide

Titanium Dioxide is one of the most widely used pigments in cosmetics because of its unique technical properties and the fact that it is universally permitted. However, in recent years, there have been some growing concerns with regard to the safe use of this pigment in cosmetics, particularly in powder form, in which various risk assessments and analytical analyses have been conducted. As a result, there have been some regulatory updates on Titanium Dioxide in the European Union and United States. These are worth mentioning briefly in this chapter, considering its importance in the cosmetics world.

CMR Classification (Category 2 by Inhalation): As per Commission Regulation (EU) 2020/217, Titanium Dioxide (in powder form containing 1% or more of particles with aerodynamic diameter of ≤ 10 μm) will be classified as a CMR category 2 substance (by inhalation) from October 1, 2021 (European Commission 2019). Following this classification, an opinion from the Scientific Committee on Consumer Safety (SCCS) would be required to determine whether Titanium Dioxide is still safe enough to be used in cosmetic products. In October 2020, an SCCS opinion on Titanium Dioxide was adopted, in which Titanium Dioxide was considered to be safe for general consumers when used in face products in powder form up to a maximum concentration of 25% and in hair products in aerosol dispenser (spray) form up to a maximum concentration of 1.4% (European Commission 2020). The use of pigmentary Titanium Dioxide as a colorant in accordance with Annex IV to Regulation (EC) No 1223/2009 should also be allowed, with specific use restrictions being added to Annex III (List of Restricted Substances) as a result of this SCCS opinion.

E171 Specification Update: Apart from its use in cosmetics, Titanium Dioxide has also historically been used as a food additive in accordance with the EU E171 specification. There is a recent update to the E171 specification as a result of the EFSA Scientific Opinion (European Food Safety Authority 2019), which was adopted in June 2019 (European Commission

2019). According to Annex IV of Regulation (EC) No 1223/2009, the Titanium Dioxide entry includes a dynamic reference to the E171 purity criteria outlined in Commission Regulation (EU) No 231/2012 of Regulation (EC) No 1333/2008 (European Commission 2009). Consequently, the proposed E171 specification amendment may also have an impact on the cosmetics industry for its use as a colorant. However, it has to be noted that the proposed E171 specification is based on the EFSA assessment and opinion on Titanium Dioxide as a food additive, and not as a cosmetic colorant; thus, the outcome of the modification on the use of Titanium Dioxide in cosmetics is being reviewed. The cosmetic industry, through multiple channels, is in discussion to address this topic, in order to ensure continued compliance of all Titanium Dioxide-containing ingredients in cosmetic applications.

Particle Size: The current definition of a "nanomaterial" as per Regulation (EC) No 1223/2009 (European Commission 2009) is not consistent with that of Commission Recommendation (2011/696/EU) (European Commission 2011), which creates room for interpretation. Further clarification with regard to the definition and measurement methodology is required to avoid confusion and inconsistency. With the evolution of analytical techniques and sample preparation methods in recent years, the particle size of some cosmetic pigments has been under increasing scrutiny, including that of Titanium Dioxide. The use of Titanium Dioxide in nano form is permitted as a UV-filter in sunscreen products, provided that it meets the conditions set out as per Commission Regulation (EU) 2016/1143 (European Commission 2016). However, the nano form of Titanium Dioxide is currently not authorized as a colorant according to Annex IV to Regulation (EC) No 1223/2009 (European Commission 2009).

United States (Continued)

California Proposition 65: In 2011, the California Office of Environmental Health Hazard Assessment (OEHHA) added Titanium Dioxide (airborne, unbound particles of respirable size) to the list of chemicals known to the State of California to cause cancer for purposes of the Safe Drinking Water and Toxic Enforcement Act of 1986 (Proposition 65). The listing, however, does not cover Titanium Dioxide when it remains bound within a product matrix (OEHHA 2011). It is worth pointing out that the use of Titanium Dioxide pigments, which are defined as "unbound particles of respirable size," in cosmetic formulations does not necessarily equate to their being unbound in the final cosmetic products exposed to consumers. The businesses involved in the manufacture of final products are responsible for conducting risk and exposure assessments to determine whether a Proposition 65 labelling is required on the packaging of products supplied to California. While the regulatory situation portrayed here is accurate at the time of writing, regulations change on a regular basis. Therefore, the information presented here may have been revised by the various regulatory bodies in the intervening time between this writing and your reading.

References

European Commission. (2009). Regulation (EC) No. 1223/2009 of the European Parliament and of the Council of 30 Nov 2009. http://eur-lex.europa.eu/LexUriServ/LexUriServ.do?uri=OJ:L:2009:342:0059:0209:en:PDF (accessed September 28, 2020).

European Commission. (2011). Commission Recommendation of 18 October 2011 on the definition of nanomaterial. https://eur-lex.europa.eu/eli/reco/2011/696/oj (accessed September 28, 2020).

European Commission. (2016). Commission Regulation (EU) 2016/1143 of 13 July 2016 amending Annex VI to Regulation (EC) No 1223/2009 of the European Parliament and of the Council on cosmetic products. https://eur-lex.europa.eu/eli/reg/2016/1143/oj (accessed September 28, 2020).

European Commission. (2019). Commission Delegated Regulation (EU) 2020/217 of 4 October 2019 amending, for the purposes of its adaptation to technical and scientific progress, Regulation (EC) No 1272/2008 of the European Parliament and of the Council on classification, labelling and packaging of substances and mixtures and correcting that Regulation. https://eur-lex.europa.eu/eli/reg_del/2020/217/oj (accessed September 28, 2020).

European Commission. (2020). Opinion on Titanium dioxide (TiO2) used in cosmetic products that lead to exposure by inhalation. https://ec.europa.eu/health/sites/health/files/scientific_committees/consumer_safety/docs/sccs_o_238.pdf (accessed September 28, 2020).

European Food Safety Authority. (2019). Scientific opinion on the proposed amendment to the EU specifications for titanium dioxide (E171) with respect to the inclusion of additional paramenters related to its particle size distribution. https://efsa.onlinelibrary.wiley.com/doi/epdf/10.2903/j.efsa.2019.5760 (accessed September 28, 2020).

FDA. (2018a). Code of Federal Regulations Title 21, Parts 1–99, revised as of April 1, 2018. https://www.accessdata.fda.gov/scripts/cdrh/cfdocs/cfcfr/CFRSearch.cfm?CFRPart=1–91 (accessed September 28, 2020)

FDA. (2018b). Code of Federal Regulations Title 21, Part 80, revised as of April 1, 2018. https://www.accessdata.fda.gov/scripts/cdrh/cfdocs/cfcfr/CFRSearch.cfm?CFRPart=80 (accessed September 28, 2020).

FDA. (2018c). Code of Federal Regulations Title 21, Part 73, revised as of April 1, 2018.   https://www.accessdata.fda.gov/scripts/cdrh/cfdocs/cfcfr/CFRSearch.cfm?CFRPart=73 (accessed September 28, 2020).

FDA. (2018d). Code of Federal Regulations Title 21, Part 74, revised as of April 1, 2018.   https://www.accessdata.fda.gov/scripts/cdrh/cfdocs/cfcfr/CFRSearch.cfm?CFRPart=73 (accessed September 28, 2020).

FDA. (2018e). Code of Federal Regulations Title 21, Part 82, revised as of April 1, 2018.   https://www.accessdata.fda.gov/scripts/cdrh/cfdocs/cfcfr/CFRSearch.cfm?CFRPart=73 (accessed September 28, 2020).

FDA. (2018f). Report on the Certification of Color Additives – Fiscal Year 2018. https://wayback.archive-it.org/7993/20190208042749/https://www.fda.gov/ForIndustry/ColorAdditives/ColorCertification/ColorCertificationReports/ucm622473.htm   (accessed   September   28, 2020).

Hochheiser, H.S. (1982). *Synthetic Food Colors in the United States: A History Under Regulation*. Ann Arbor, MI: University Microfilms International.

MHLW. (2003). Ordinance No. 30/1966; Ordinance to Regulate Coal-Tar Colors Permitted for Use in Drugs, Quasi-drugs and Cosmetics (as amended by Ordinance No. 126/2003). Ministry of Health, Labour and Welfare, Tokyo, Japan.

NMPA. (2015). Safety and Technical Standards for Cosmetics. National Medical Product Administration, People's Republic of China, Beijing, China (2015 edition) http://www.sesec.eu/app/uploads/2016/02/Cosmetics-Safety-and-Technical-Standards-2015-Version-Foreword-and-summary.pdf (accessed September 28, 2020).

OEHHA. (2011). Titanium dioxide (airborne, unbound particles of respirable size). California Office of Environmental Health Hazard Assessment, Sacremento, CA, USA. https://oehha.ca.gov/proposition-65/chemicals/titanium-dioxide-airborne-unbound-particles-respirable-size (accessed September 28, 2020).

Rosholt, A.P., Esq (2007). *International Color Handbook*, 4e. Washington, DC: Personal Care Products Council.

USDA. (n.d.). United States Department of Agriculture. https://www.fsis.usda.gov/wps/portal/fsis/home (accessed September 28, 2020).

Yakuji Nippo (1989). *Principles of Cosmetic Licensing in Japan*, 2e. Tokyo: Yakuji Nippo.

Chapter 3
Color Selection – Stability

Once the regulatory requirements for the market where a cosmetic or toiletry product will be sold are satisfied, it must be determined that the colorants used in the formulation will be stable. Stability bears several specific criteria. First, the colorants must not react with other chemicals in the formula. Second, they must be stable under the conditions to which they will be subjected during the manufacturing process, including, but not limited to, temperature, time, concentration, and pH. Third, they must be stable under the conditions to which they will be exposed during display and use, including, but not limited to, light, humidity, and changes in temperature.

As mentioned in the Introduction, only a small number of approved color additives are used on a regular basis in the formulation of cosmetics and toiletries. Therefore, to simplify the examination of stability, only these important, frequently used colors will be discussed.

In order to study the stability of color additives, it is first necessary to divide them into different chemical classes. The two major categories are the synthetic organic colorants, or as the United States Food and Drug Administration (FDA) defines them, the **certifiable colors**, and the FDA-designated **exempt colors**.

Coloring the Cosmetic World: Using Pigments in Decorative Cosmetic Formulations,
Second Edition. Edwin B. Faulkner. Edited by Jane C. Hollenberg.
© 2021 John Wiley & Sons Ltd. Published 2021 by John Wiley & Sons Ltd.

Certifiable (Synthetic Organic) Colorants

The certifiable colors can be broadly separated into five subcategories based on their chemical structures:

1. Nonsoluble azo colors.
2. Soluble azo colors.
3. Slightly soluble azo colors.
4. Soluble non-azo colors.
5. Miscellaneous colors.

The general physical properties of these organic colors are as follows:

- They are typically hydrophobic and lipophilic, so they will wet preferentially in oil.
- They exhibit good to poor light stability.
- Their chemical stability ranges from good to poor.
- They exhibit moderate to poor bleed resistance.
- They are difficult to disperse.
- They exhibit bright, clean, vibrant shades.
- They have excellent tinctorial properties.

The colorants in each of the subcategories will exhibit similar stability characteristics.

Non-soluble Azo Colors

The most common colorant in this category is D&C Red No. 36. A photograph of the pure pigment is shown in Figure 3.1a, while its chemical structure is shown in Figure 3.1b. As can clearly be seen, the double-bonded nitrogen atoms form the azo linkage, which is one of the earliest, and still most important, means of producing colored compounds. Please note: this N=N bond defines an *azo* grouping. (Interestingly, the name "azo" comes from *azote*, the French name for nitrogen, which is derived from the Greek *a* [not] + *zoe* [to live].) These double-bonded nitrogen atoms, combining two aromatic structures, set up a resonance in the molecule, which results in the color as it is seen by the eyes. "Azo colors," as color additives with this type of structure are called, are the most important class of synthetic organic colorants used in cosmetic and toiletry products.

D&C Red No. 36 contains no water solubilizing groups, such as SO_3 or COOH, to give it water solubility. At the same time, though completely

(a)

(b)

Figure 3.1   D&C Red No. 36: (a) photograph; (b) chemical structure.
Source: Courtesy of Sun Chemical Corp.

organic, it is not soluble in normal organic oils or hydrocarbons because of its structure or inner charges.

Colors of this type are hydrophobic and bleed resistant to water and alkali, and exhibit very good light stability. They do have a slight tendency to bleed in oil or organic solvents. Because of this, they should not be used in products that contain strong solvents, such as nail lacquers. One other issue with D&C Red No. 36 is a tendency to bleed slightly in the oils and waxes used in lipstick. This slight solubility causes it to crystallize and darken upon continued reheating of the bulk; however, when used in a single-heat bulk lipstick, crystallization normally will not be a problem.

Soluble Azo Colors

A typical soluble azo color is D&C Red No. 33. A photograph of the pure pigment is shown in Figure 3.2a, while its chemical structure is shown in Figure 3.2b. Once again, the double-bonded nitrogen atoms connecting the aromatic rings form the azo group. The two SO_3 groups make this colorant quite water-soluble.

(a)

(b)

Figure 3.2   D&C Red No. 33: (a) photograph; (b) chemical structure.
Source: Courtesy of Sun Chemical Corp.

All of the color additives in this group are water-soluble dyes that are used in this form to color toiletry products. They are precipitated as classical chemical lakes to form the pigments that are used in decorative cosmetics. The dyes are quite resistant to both acid and alkali, and have reasonable light stability. However, their pigments (lakes) will tend to bleed in water and are not stable in either strong acid or strong alkali. Under these conditions, the lakes break down and release some of the soluble dye, which is the source of the bleed.

The lakes of these colorants are stable in oils, waxes, and aromatic solvents. In addition, they exhibit good heat stability up to 100 °C. As a result of these properties, these lakes are widely used in lipsticks and nail polish. Some of the other important colorants in the soluble azo category are FD&C Yellow No. 5, FD&C Yellow No. 6, and FD&C Red No. 40. Photographs of the pure pigments are shown in Figure 3.3a–c. Another soluble azo colorant is D&C Orange 4, which is restricted to external only use in the US. The lake has not been manufactured recently due to the limited potential volume of an orange that could not be utilized in lip products.

Figure 3.3 (a) FD&C Yellow No. 5. (b) FD&C Yellow No. 6. (c) FD&C Red No. 40.
Source: Courtesy of Sun Chemical Corp.

Slightly Soluble Azo Colors

This class of colorants is the most important of the synthetic organic colors used in decorative cosmetics. All of its members are toners or reduced toners. A typical member of the category is D&C Red No. 6. A photograph of the pure pigment is shown in Figure 3.4a, while its chemical structure is shown in Figure 3.4b. This color additive, even though it has both an SO_3 group and a COOH group, is not very water-soluble, because precipitation as a metal salt affects its polar nature. As discussed in Chapter 2, this color is important as a universal yellow shade red for the US, EU, and Japanese markets. Even more important in the United States and the European Union are the barium salt of D&C Red No. 6 (Figure 3.5a) and the calcium salt of the molecule, which is D&C Red No. 7 (Figure 3.5b). These colorants are so important because they provide excellent color value and economics (see Chapter 4). Both exhibit good light stability, are heat-stable to approximately 105 °C, and are resistant to bleed in oils and solvents. This makes them very useful in powders, lipsticks, and nail polish. Unfortunately, they do not have the same stability properties in aqueous-based products. In strong alkali, they turn

(a)

(b)

Figure 3.4   D&C Red No. 6: (a) photograph; (b) chemical structure.
Source: Courtesy of Sun Chemical Corp.

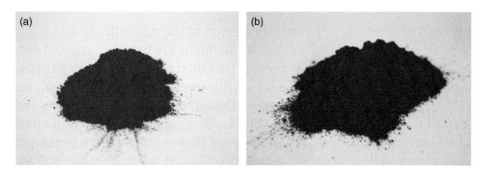

Figure 3.5   (a) Barium salt of D&C Red No. 6. (b) D&C Red No. 7.
Source: Courtesy of Sun Chemical Corp.

yellow, and in strong acid, they turn bluish-red, limiting their usability in these types of systems.

The other two colors worth noting in this category are D&C Red No. 31 and D&C Red No. 34 Ca Lake. A photograph of D&C Red No. 34 Ca Lake pigment is shown in Figure 3.6. Though they have the same stability properties as the other colorants in the category, they are more limited in their use due

Figure 3.6   D&C Red No. 34 Ca Lake.
Source: Courtesy of Sun Chemical Corp.

to FDA regulatory restrictions. Neither can be used in mucous membrane contact cosmetics (i.e., lipsticks). Additionally, D&C Red No. 34 Ca Lake has a slight tendency to bleed in alcohol, further limiting its usefulness to nail lacquer, where its vivid violet color is an asset.

Soluble Non-Azo Colors

Most of the important colorants in this category are in the Fluorescein family. The name derives from the fact that they can fluoresce (see Chapter 10). D&C Red No. 22, the sodium salt of tetrabromofluorescein, is a typical example. A photograph of the pure pigment is shown in Figure 3.7a, while its chemical structure is shown in Figure 3.7b. The other color additives in this family are listed in Table 3.1. Members of the Fluorescein family of dyes can act as pH indicators, changing colors in response to acid/base. All will stain the skin, which led to their use in the development of long-wearing lipsticks. The Fluorescein pigments are classical chemical lakes that are used for their unique clean shades, in spite of their poor stability properties. From a temperature standpoint, they can only be kept at 100 °C for short periods of time. They bleed slightly in aqueous systems and rather heavily in solvents, making them completely unsuitable for nail polish use. Finally, they are quite fugitive to light – natural, incandescent, and fluorescent. In fact, in order to use these lakes in today's clear-top packaging, ultraviolet (UV) absorbers must be incorporated into the plastic packaging. Photographs of the lakes of D&C Orange No. 5, D&C Red No. 21, and D&C Red No. 27 are shown in Figure 3.8a–c.

(a)

(b)

Figure 3.7   D&C Red No. 22: (a) photograph; (b) chemical structure.
Source: Courtesy of Sun Chemical Corp.

Table 3.1   Fluorescein Family Colorants.

FDA name	Chemical description	Form
D&C Orange No. 5	Dibromofluorescein	Sodium salt
D&C Red No. 21	Tetrabromofluorescein	Free acid
D&C Red No. 22	Tetrabromofluorescein	Sodium salt
D&C Red No. 27	Tetrabromo-tetrachlorofluorescein	Free acid
D&C Red No. 28	Tetrabromo-tetrachlorofluorescein	Sodium salt

Miscellaneous Colors

The most useful color additive in this category for decorative cosmetics is D&C Red No. 30, an indigoid vat colorant (i.e., characterized by the same chromophore as indigo). A photograph of the pure pigment is shown in Figure 3.9a, while its chemical structure is shown in Figure 3.9b. It is an excellent color for use in most types of decorative cosmetics. As an industrial color, it is used extensively in textile printing inks, so it has very good stability

Figure 3.8 (a) D&C Orange No. 5. (b) D&C Red No. 21. (c) D&C Red No. 27.
Source: Courtesy of Sun Chemical Corp.

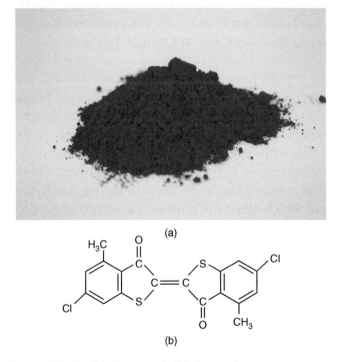

Figure 3.9 D&C Red No. 30: (a) photograph; (b) chemical structure.
Source: Courtesy of Sun Chemical Corp.

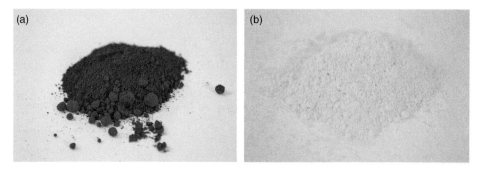

Figure 3.10   (a) FD&C Blue No. 1. (b) D&C Yellow No. 10.
Source: Courtesy of Sun Chemical Corp.

to light, heat (100 °C range), acids, alkalis, and most organic solvents. The one stability problem is its slight tendency to bleed in acetone. As a result of this characteristic, D&C Red No. 30 finds only limited use in nail polish.

Other important color additives in the miscellaneous category are the lakes FD&C Blue No. 1 and D&C Yellow No. 10. Photographs of them are shown in Figure 3.10a and Figure 3.10b, respectively. FD&C Blue No. 1 Al Lake is a triphenylmethane colorant, while D&C Yellow No. 10 Al Lake is a quinoline colorant. Like the Fluorescein color additive family, these colors have very pure, clean shades, but unfortunately, they also exhibit only fair stability characteristics. Their lightfastness is poor, they bleed in solvents, and their lakes tend to have a slight water bleed. Finally, they cannot be exposed to temperatures at or in excess of 100 °C for long periods of time.

As a means of summarizing what we have learned to this point, Table 3.2 lists all US approved certifiable primary colorants alongside details of their stability properties with regard to water, solvents, and light.

Now, before moving on to the exempt color additives, a review of the US listed substrata is in order. There are 10 substrates permitted for use in making lakes, five of which affect the physical properties of the color additives, and five of which do not.

The ones in the first category are:

Aluminum Hydroxide: Commonly called aluminum hydrate, aluminum hydroxide is the most common permitted substrate, as it is used to make all of the FD&C lakes and the majority of the true chemical lakes of the D&C colors. It affects the colorant by rendering the water-soluble dye insoluble as the dye is absorbed on to it.

Table 3.2   Stability of US Certifiable Colors (Zuckerman and Senackerib 1979).

Name	H$_2$O	Glycerol	Methanol	Ethanol	Petroleum jelly
FD&C Green No. 3	S	S	S	M	I
FD&C Yellow No. 5	S	S	SS	SS	IE
FD&C Yellow No. 6	S	S	S	SS	I
FD&C Red No. 4	S	S	SS	SS	IE
FD&C Red No. 40	S	S	S	SS	I
FD&C Blue No. 1	S	S	S	S	C
D&C Green No. 5	S	S	S	SS	IE
D&C Green No. 6	I	Ia	SS	SS	M
D&C Green No. 8	SF	SSF	SSF	SSF	Ia
D&C Yellow No. 7	IBF	SSF	SSF	SS	D
D&C Yellow No. 8	SF	SF	SF	M	IE
D&C Yellow No. 10	S	S	M	SS	I
D&C Yellow No. 11	I	SS	S	S	S
D&C Red No. 6	S	S	SS	Ia	I
D&C Red No. 7	I	D	Ia	Ia	D
D&C Red No. 17	I	SS	SS-M	SS	S
D&C Red No. 21	IBF	Da	SS	SS	D
D&C Red No. 22	SF	SF	SF	SF	IE
D&C Red No. 27	IB	Da	SS	SS	D
D&C Red No. 28	S	S	S	S	IE
D&C Red No. 30	IU	D	I	I	I
D&C Red No. 31	M	SS	SS	SS	I
D&C Red No. 33	S	S	SS	SS	I
D&C Red No. 34	I	I	Ia	I	D
D&C Red No. 36	I	D	Ia	Ia	D
D&C Orange No. 4	S	S	S	M	IE
D&C Orange No. 5	IB	SS	S	M	D
D&C Blue No. 4	S	S	S	S	C
D&C Brown No. 1	S	S	S	SS	IE
D&C Violet No. 2	I	Ia	SS	SS	S
Ext. D&C Violet No. 2	S	S	SS	SS	I
Ext. D&C Yellow No. 7	S	S	M	SS	I

Name	Toluene	Stearic acid	Oleic acid	Mineral oil	Mineral wax
FD&C Green No. 3	I	I	I	I	I
FD&C Yellow No. 5	IE	IE	IE	IE	IE
FD&C Yellow No. 6	I	I	I	I	I
FD&C Red No. 4	I	IE	IE	IE	IE
FD&C Red No. 40	I	I	I	I	I
FD&C Blue No. 1	I	C	C	C	C
D&C Green No. 5	I	IE	IE	IE	IEW
D&C Green No. 6	S	M	M	M	M

(*continued*)

Table 3.2   (*continued*)

Name	Toluene	Stearic acid	Oleic acid	Mineral oil	Mineral wax
D&C Green No. 8	I	Ia	Ia	I	I
D&C Yellow No. 7	I	D	D	D	D
D&C Yellow No. 8	I	IE	IE	IE	IE
D&C Yellow No. 10	I	I	I	I	I
D&C Yellow No. 11	S	S	S	S	S
D&C Red No. 6	I	I	I	I	I
D&C Red No. 7	I	D	D	D	D
D&C Red No. 17	S	S	S	S	S
D&C Red No. 21	I	D	D	D	D
D&C Red No. 22	I	IE	IE	IE	IE
D&C Red No. 27	I	D	D	D	D
D&C Red No. 28	I	IE	IE	IE	IE
D&C Red No. 30	Ia	D	D	D	D
D&C Red No. 31	I	I	I	I	I
D&C Red No. 33	I	I	I	I	I
D&C Red No. 34	I	D	D	D	D
D&C Red No. 36	I	D	D	D	D
D&C Orange No. 4	I	IE	IE	IE	IE
D&C Orange No. 5	I	D	D	D	D
D&C Blue No. 4	I	C	C	C	C
D&C Brown No. 1	I	IE	IE	IE	IE
D&C Violet No. 2	S	S	S	S	SW
Ext. D&C Violet No. 2	I	I	I	I	I
Ext. D&C Yellow No. 7	I	I	I	I	I

Name	Ethyl ether	Acetone	Butyl acetate	Light	10% acetic acid	10% HCl	10% NaOH	0.9% Physiol. saline
FD&C Green No. 3	I	I	I	3	5	5	2b	6
FD&C Yellow No. 5	I	I	I	5	5	5	5	6
FD&C Yellow No. 6	I	I	I	3	5	5	5	6
FD&C Red No. 4	I	I	I	6	6	5	5	6
FD&C Red No. 40	I	I	I	3	5	5	5	6
FD&C Blue No. 1	I	Ia	I	3	5	4g	4	6
D&C Green No. 5	I	SS	I	5	5	5	5	5
D&C Green No. 6	SS	SS	S	4	5L	51	61	I
D&C Green No. 8	Ia	Ia	Ia	2	I	I	5	6
D&C Yellow No. 7	SS*	S	I	2	I	I	S6	I
D&C Yellow No. 8	la	kIa	I	3	3p	3p	6	6
D&C Yellow No.10	Ia	SS	I	3	5	5	4r	6
D&C Yellow No. 11	S	S	S	2	I	51	Iw	I
D&C Red No. 6	I	Ia	I	5	5	4	4d	6

(*continued*)

Table 3.2 (*continued*)

Name	Ethyl ether	Acetone	Butyl acetate	Light	10% acetic acid	10% HCl	10% NaOH	0.9% Physiol. saline
D&C Red No. 7	I	Ia	I	6	5I	4I	5I	I
D&C Red No. 17	SS	SS	M	3	5L	4Id	5I	I
D&C Red No. 21	M*	S	I	2	3I	3I	5Sr	1
D&C Red No. 22	Ia	SS	I	2	2py	lpy	5	6
D&C Red No. 27	Ia	SS	I	2	3I	3	5Sr	I
D&C Red No. 28	Ia	SS	I	3	2p	4p	6	6
D&C Red No. 30	Ia	Ia	Ia	6	7I	I	6IU	I
D&C Red No. 31	I	Ia	Ia	5	5	4	5	6
D&C Red No. 33	I	I	I	5	6	3z	5	6
D&C Red No. 34	l	D	D	4	5I	4	4I	I
D&C Red No. 36	I	Ia	D	6	6I	4d	4d	I
D&C Orange No. 4	l	Ia	I	5	5	5	2m	6
D&C Orange No. 5	M	S	I	2	4aI	4I	Sr	I
D&C Blue No. 4	I	Ia	I	3	5	5	4	6
D&C Brown No. 1	SS	SS	l	3	5	5	6sIy	6
D&C Violet No. 2	SS	SS	S	4	5I	5I	5I	6I
Ext. D&C Violet No. 2	I	SS	I	5	5	5	5	6
Ext. D&C Yellow No. 7	I	M	I	4	5	5	5	6

Name	5% FeSo$_4$	5% Alum	Oxidizing agents	Reducing agents
FD&C Green No. 3	3v	4	2	1
FD&C Yellow No. 5	d	4	3	1
FD&C Yellow No. 6	4	4	3	1
FD&C Red No. 4	p	p	3	1
FD&C Red No. 40	4	4	3	1
FD&C Blue No. 1	4r	4	2	1
D&C Green No. 5	4	4	3	2
D&C Green No. 6	I	I	3	2
D&C Green No. 8	4d	4d	3	3
D&C Yellow No. 7	I	I	3	3
D&C Yellow No. 8	z-p	p	3	3
D&C Yellow No. 10	z	4	2	5
D &C Yellow No. 11	I	I	2	5
D&C Red No. 6	p	p	3	1
D&C Red No. 7	4Id	4I	3	1
D&C Red No. 17	4Id	4I	3	1
D&C Red No. 21	Id	4I	4	4
D&C Red No. 22	3d	2y	4	4
D&C Red No. 27	I	I	4	4
D&C Red No. 28	z	p	4	4
D&C Red No. 30	I	I	5	u

(*continued*)

Table 3.2   (*continued*)

Name	5% FeSo$_4$	5% Alum	Oxidizing agents	Reducing agents
D&C Red No. 31	p	p	3	1
D&C Red No. 33	4	4	3	1
D&C Red No. 34	I	I	3	1
D&C Red No. 36	4d	4	3	1
D&C Orange No. 4	J-p	J-p	3	3
D&C Orange No. 5	I	I	3	3
D&C Blue No. 4	4	4	2	1
D&C Brown No. 1	p	p	3	1
D&C Violet No. 2	4I	4I	2	1
Ext. D&C Violet No. 2	4z	4	3	2
Ext. D&C Yellow No. 7	zd	4	3	3

a – Very sparingly soluble; may bleed or stain.
B – Insoluble in water, soluble in aqueous alkaline solution.
b – Turns much bluer in hue.
C – Practically insoluble, but useful in nearly neutral or slightly acid emulsions.
c – At 25 °C.
D – Practically insoluble, but may be dispersed by grinding and homogenizing; solid media (waxes) should be softened or melted before or during grinding.
d – Hue becomes duller or darker.
E – Practically insoluble in fatty acids, oils, or waxes, but useful in coloring slightly alkaline aqueous emulsions.
F – Solution usually fluorescent.
G – Soluble or dispersible in oils and waxes in the presence of 10–25% of a fatty acid.
g – Turns much greener in hue.
I – Insoluble.
J – Tends to thicken or gel the solution.
k – Turns brownish in hue.
L – Turns orange in hue.
M – Moderately soluble (<1%).
m – Turns scarlet in hue.
p – Dye precipitated as a heavy metal salt or colored acid.
r – Turns redder in shade.
S – Dissolves (solubility >1%).
SS – Sparingly soluble (<0.25%).
U – In alkaline reducing vats, a soluble leuco compound forms.
v – Turns violet in hue.
W – Not fast to prolonged storage in some waxes.
w – Becomes tinctorially weaker.
x – Turns yellow in hue.
y – Turns yellower in hue.
z – Hazy or cloudy.
* – practically colorless.
1 – very poor fastness.
2 – Poor fastness.
3 – Fair fastness.
4 – Moderate fastness.
5 – Good fastness.
6 – Very good fastness.
7 – Excellent fastness.

Gloss White: No longer used in commercial products, gloss white is a mixture of aluminum hydroxide and barium sulfate. It was common in the days when aluminum hydroxide was made in situ as part of the laking process. The effect on the color was the same as that of neat aluminum hydroxide.

Rosin: Primarily abietic acid. The most interesting of the substrata, it is found exclusively in the toners, where its purpose is to increase darkness and transparency. Toners made without it are light and opaque, while those with rosin levels of 12–15% are quite dark and transparent in mass tone. As a general rule, with the other substrates, their ability to reduce strength is directly proportional to the amount used in the lake. Rosin, unlike the others, will have no effect on the strength of a toner when used at levels under 10%. Toners with rosin contents >15% are not normally used in nail polish as they can cause gelling of the polish in the bottle.

Aluminum Benzoate: Performs almost identically to aluminum hydroxide; the difference is that it produces lakes with a higher degree of opacity. Additionally, it is important to note, there is a low probability of bacterial growth not ordinarily experienced when using the aluminum hydroxide.

Calcium Carbonate: Like gloss white, calcium carbonate is no longer used commercially, because in aqueous systems with a pH below 7, there is the possibility of decomposition generating carbon dioxide in the finished cosmetic product.

Now, on to the substrata that do *not* affect the physical properties of lakes. While, again, physical properties are unaffected, these five substrata do affect tinctorial value, as all of them act as diluents on a linear basis. They are as follows:

Barium Sulfate: Commonly called blanc fixe, it is the most common substrate used to dilute toners. The reason for its popularity is that it will not affect the opacity of the color to any great extent, and as it is not highly aggregated, it will help render the toners more dispersible.

Talc: Has been used almost exclusively to extend toners that are designed for use in powder products, where they are very compatible with the other ingredients in those products and, like barium sulfate, will help with dispersibility. Talc lakes are not normally used in lipstick formulas because they will negatively impact spreadability and adhesion, cause a chalky look on the lips, and soften stick structure. It should be noted that, as of this writing, use of talc in personal care products may be expected to virtually disappear in many countries in the near future due to concerns over alleged carcinogenicity.

Clay: Can be used interchangeably with barium sulfate where there are concerns about barium content in the finished product, whether soluble or insoluble.

Zinc Oxide: Like the other substrata in this category, zinc oxide has a linear impact on the tinctorial properties of lakes. It is not widely used because of its tendency to make the color look washed out. It should be noted that zinc oxide is both a US listed exempt color additive and a listed substrate.

Titanium Dioxide: Functions similarly to zinc oxide but provides even greater opacity. It is not widely used because of its tendency to make the color look "washed out" and too opaque. Also, like zinc oxide, titanium dioxide is both a US listed exempt color additive and a listed substrate.

Exempt Colors

The second general class of color additives is exempt colors, including both inorganic and nonsynthetic organic colors. The "major" exempt colors, those most often used, are all pigments, and therefore find very little use in toiletry products. The stability of the exempt colors will be reviewed based on their chemical structures. The general physical properties of the inorganic colors are as follows:

- They are generally somewhat hydrophilic so will wet preferentially in water.
- They exhibit excellent light stability.
- Their chemical stability is generally good.
- They exhibit excellent bleed resistance.
- With one exception (see later), they are easy to disperse.
- They exhibit dull, dirty shades.
- They have moderate tinctorial properties.

Exempt colors break down into a number of subcategories, as follows.

Iron Oxides

From a chemical stability standpoint, the iron oxides are almost ideal colorants for cosmetic products. They are relatively easy to disperse, while being inert to water, acids, alkalis, and solvents. They are very opaque and exhibit excellent light stability. Variations in molecular structure produce a number of shades. Photographs of yellow, red, and black iron oxide are shown in Figure 3.11a–c.

Figure 3.11   (a) Yellow iron oxide. (b) Red iron oxide. (c) Black iron oxide.
Source: Courtesy of Sun Chemical Corp.

There is one note of caution with regard to the use of the iron oxides, and that relates to heat. The yellow is a hydrated ferric oxide ($Fe_2O_3 \cdot H_2O$), while the red is the anhydrous version (Fe_2O_3). The red is produced by calcination of the yellow at about 800 °C, so it is very heat-stable in the relative temperature range where cosmetic products are produced. The yellow, however, will shed some of its water of hydration at temperatures as low as 125–150 °C, causing some shift toward red that is visibly evident to the naked eye.

The black iron oxide, Fe_3O_4, is a mixture of ferric and ferrous oxides. Like the yellow, it will change color, becoming redder, at temperatures of 125–150 °C. This oxidation to Fe_2O_3 is exothermic and self-perpetuating, leading to classification of Fe_3O_4, as a self-heating solid for shipping purposes, hazardous in pack sizes of larger than 25 kg. Black iron oxide is also magnetic and will coat iron or mild steel containers and agitators used in processing.

Ferric Ammonium Ferrocyanide

Ferric ammonium ferrocyanide, also known as iron blue, is an iron ammonium compound, with the structure $Fe(NH_4)Fe(CN)_6$. A photograph of

Figure 3.12   Ferric ammonium ferrocyanide (iron blue).
Source: Courtesy of Sun Chemical Corp.

the dry powder is shown in Figure 3.12. It is sometimes confused with the water-soluble ammonium salt (CI7520), which is not approved for cosmetic use in the United States, the European Union, or Japan. Ferric ammonium ferrocyanide is very stable to heat and light and has good resistance to solvents. In aqueous systems, it is stable in acid, but decomposes at pH levels above 7, causing a loss of color. Iron blue is the pigment previously alluded to that is quite difficult to disperse.

Chromium Greens

Chromium oxide green (Figure 3.13a) and chromium hydroxide green (Figure 3.13b) are typical mineral pigments with excellent light stability, heat stability, and bleed resistance. The hydrated version is much brighter, cleaner, and bluer than the anhydrous one. Both colors are also very stable in alkalis and acids, as well as in solvents. In some formulations, the hydrated green will react with some perfume oils, resulting in an off odor.

Ultramarines

The ultramarines are truly excellent colors, for several reasons. They are available in a wide range of shades, including violet (Figure 3.14a), pink (Figure 3.14b), and the most popular, blue (Figure 3.14c). All are complex sodium aluminum sulfosilicates, having the general structure Na(AlSiO)S. The ultramarines are excellent colors due to their stability to light, heat, and most solvents. They do have one drawback, however: while stable to alkalis,

Figure 3.13 (a) Chromium oxide green. (b) Chromium hydroxide green.
Source: Courtesy of Sun Chemical Corp.

Figure 3.14 (a) Ultramarine violet. (b) Ultramarine pink. (c) Ultramarine blue.
Source: Courtesy of Sun Chemical Corp.

they decompose in the presence of acid and evolve hydrogen sulfide (H_2S), the smell of rotten eggs – not a very pleasant fragrance in cosmetic products.

Manganese Violet

Manganese violet is a manganese ammonium pyrophosphate complex ($Mn(III)NH_4P_2O_7$), available in both "red" (Figure 3.15a) and "blue"

Figure 3.15   (a) Manganese violet red shade. (b) Manganese violet blue shade.
Source: Courtesy of Sun Chemical Corp.

(Figure 3.15b) shades that are unusually bright and clean for an inorganic pigment. It has excellent light stability, is resistant to organic solvents, and has good heat stability in the temperature range encountered during cosmetic product manufacture. It is stable to acids, but will decompose in neutral to alkaline systems, in which the color will disappear or turn black, depending on pH.

Titanium Dioxide

Titanium dioxide (TiO_2), shown in Figure 3.16, is the primary white pigment used to provide opacity and coverage in color cosmetics. Based on the fact that pigmentary titanium dioxide is inert in cosmetic formulations, the cosmetic chemist has wide latitude in the types of products in which it can be used. It is insoluble in water, resistant to acids and alkalis, and stable in organic solvents. In all cosmetic products, light and heat stability are excellent. Pigmentary titanium dioxide is permitted for general cosmetic use by all countries, but, as with all colorants, current regulations for all regions in which a product will be marketed should be consulted prior to use. (See Section "Regulatory Changes on the Horizon in the United States, European Union, China, and Japan" for specific regulatory concerns regarding titanium dioxide.)

Zinc Oxide

Zinc Oxide (ZnO) is the other white pigment generally approved for worldwide cosmetic use. It is less covering than titanium dioxide and has a slightly yellowish tone. The main deterrent to its greater usage in aqueous formulations is a slight solubility in water, which is greatly increased at

Figure 3.16 Titanium dioxide.
Source: Courtesy of Sun Chemical Corp.

pH values below 5.5. Even trace amounts of solubilized Zn^{+2} ion can affect emulsion stability and cause agglomeration of other pigments. Hydrophobic surface treatments (Chapter 8) are used to prevent solubilization of zinc oxide, permitting incorporation is the oil phases of emulsions. In foundation shades for darker skin tones, zinc oxide can provide coverage with a less "ashy" appearance than the strong bluish white undertone of titanium dioxide.

Carmine

Probably the most interesting color additive used in decorative cosmetics, carmine is the aluminum lake of carminic acid, which is the processed extract of dried female Cochineal beetles. A photograph of the laked pigment is shown in Figure 3.17. The beetles are harvested predominately in Peru, though some come from the Canary Islands. For a natural colorant, carmine exhibits good stability. It is stable in most organic solvents and has fair light and heat stability (to ~60 °C). Being a chemical lake, it can be broken down in strong acids and bases, liberating free carminic acid, and is partially soluble in water.

By way of summary, Table 3.3 offers a listing of the more important exempt color additives, showing their properties.

Having explored the wide variety of colorants in this chapter, it should now be evident to the reader that colorants are not inert substances about whose stability the formulator need not be concerned. Color additives are

Figure 3.17 Carmine.
Source: Courtesy of Sun Chemical Corp.

Table 3.3 Properties of Permitted Exempt Colors.

Aluminum powder	Insoluble in all water and solvents used in cosmetics and toiletries; excellent lightfastness Can generate H_2 in aqueous systems.
Annato	Insoluble in water; somewhat soluble in alcohol and oils; soluble in alkali, precipitated by acids; poor lightfastness.
Bismuth oxychloride	Insoluble in all water and solvents used in cosmetics and toiletries; somewhat poor lightfastness.
Bronze powder	Insoluble in all water and solvents used in cosmetics and toiletries; excellent lightfastness; will tarnish under acidic conditions and in the presence of ammonia.
Caramel	Soluble in water; insoluble in organic solvents; fast to light, weak acids and weak alkalis.
Carmine	Insoluble in water; insoluble in organic solvents; will break down in strong acid and alkalis; moderate light fastness.
Carotene	Insoluble in water; soluble in oils; good lightfastness.
Chlorophyllin-copper complex	Soluble in water, alkalis; fairly good fastness to light, heat; insoluble in dilute acids, forming free acid form which is insoluble in oils.
Chromium hydroxide green	Insoluble in water; insoluble in organic solvents; insoluble in acids and alkalis; fast to light.
Chromium oxide green	Insoluble in water; insoluble in organic solvents; insoluble in acids and alkalis; fast to light.

(continued)

Table 3.3   (*continued*)

Copper powder	Insoluble in all water and solvents used in cosmetics and toiletries; excellent lightfastness will tarnish under acidic conditions and in the presence of ammonia.
Dihydroxyacetone	Soluble in water; soluble in alcohol; insoluble in ether; turns brown on application to the skin.
Ferric ammonium ferrocyanide	Insoluble in water; insoluble in organic solvents; insoluble in acids; not stable in alkalis; good lightfastness.
Ferric ferrocyanide	Insoluble in water; insoluble in organic solvents; insoluble in acids; not stable in alkalis; good lightfastness.
Iron oxides	Insoluble in water; insoluble in organic solvents; insoluble in alkalis; attacked by strong acids; excellent lightfastness.
Manganese violet	Insoluble in water; insoluble in organic solvents; stable in acids; not stable in water at neutral to alkaline pH. excellent lightfastness.
Titanium dioxide	Insoluble in water; insoluble in organic solvents; insoluble in acids and alkalis; good light stability in cosmetics.
Ultramarines	Insoluble in water; insoluble in organic solvents; not stable in acids; insoluble in alkalis; excellent lightfastness.
Zinc oxide	Insoluble in water; insoluble in organic solvents; insoluble above pH 5.5; excellent light stability.
Silver	Insoluble in all water and solvents used in cosmetics and toiletries; excellent lightfastness.

like all other ingredients in that they are subject to interactions with other chemicals. Potential interactions with other ingredients must be considered when formulating cosmetic products.

Reference

Zuckerman, S. and Senackerib, J. (1979). Colorants for foods, drugs, and cosmetics. In: *Kirk-Othmer Encyclopedia of Chemical Technology*, 3e, vol. 6, 570–573. Hoboken, NJ: Wiley.

Chapter 4

Color Selection – Color Esthetics

Introduction

O nce the first two requirements of color selection – understanding and adhering to respective countries' regulations, and selecting colorants in order to formulate products with stability – are satisfied, the cosmetic chemist can then concern him/herself with the third requirement, one that is crucial particularly in order to craft a product that is attractive to the consumer: color selection based on color esthetics.

Formulating color cosmetic and toiletry products is not unlike an activity that all children enjoy: finger painting. The joy of mixing colored paints together to create new and different shades is a satisfying experience for all children. It builds tactile abilities and engages the eyes, mind, and motor functions in creating new and interesting visual stimulation. Mixing color additives to create unique and beautiful shades provides the cosmetic chemist with a similar feeling of satisfaction, engaging the creative faculties in both pleasurable and challenging ways.

The subject of color selection based on esthetics will be covered from two aspects. The first will look at colors from the perspective of color shade, separated into organic and inorganic types. The second will focus on the selection of color additives by type of finished cosmetic product, such as lipstick, eye shadow, and nail polish.

Coloring the Cosmetic World: Using Pigments in Decorative Cosmetic Formulations,
Second Edition. Edwin B. Faulkner. Edited by Jane C. Hollenberg.
© 2021 John Wiley & Sons Ltd. Published 2021 by John Wiley & Sons Ltd.

Before We Begin ...

Before embarking on the esthetic selection journey, there are a few basic rules of color formulating that apply to all decorative cosmetic products:

Multiple Color Additives: Cosmetic and toiletry products are rarely formulated with a single color additive. There are almost always at least three used, and sometimes (with the use of effect pigments) as many as six or seven. There are two reasons for using multiple colorants. First, there are slight shade differences from batch to batch of the same color additive, so having more than one colorant in a formula allows for slight adjustments in their ratios, in order to produce a finished product with the exact same shade every time. Second, the use of several different colorants makes copying the product much more difficult for a competitor.

Color Extremes: The use of color additives with extreme differences in shade is normally avoided. The reason for this is that mixing these extremes will result in a dirty appearance in the color of the final product.

Color Combinations: There are major differences in the prices of color additives, based on their manufacturing costs and the costs of the raw materials used to produce them. In many cases, there is more than one combination of colorants that will result in a particular finished product color point. Bearing in mind the cost of the individual components, the most economical combination should always be chosen. This subject will be covered in more detail in Chapter 5.

Oil Absorption: The tendency to absorb oil from a formula varies considerably from colorant to colorant. Generally speaking, the synthetic organic colors, particularly the aluminum hydrate lakes, exhibit higher oil absorption than the inorganic colorants. This variation often requires the chemist to adjust the percentage of oil from shade to shade in a particular product line. The adjustment of oil percentage will insure that the stick hardness in lipsticks will be the same throughout the line. A similar approach also insures equal rheology among shades in pigmented liquid oil formulations and emulsions.

Dispersion: Dispersion is the process of separating the colorant agglomerates and distributing them throughout the cosmetic product. This step is necessary in order to develop the full tinctorial value from a color additive. Like oil absorption, the ease of dispersion varies from colorant to colorant. Therefore, it is strongly recommended to make single monochromatic (single-color) dispersions and to vary the energy input according to the amount needed to insure full color development.

Differences Within Color Families: Up to this point, all colorants discussed in this book have been presented as though there were only one version of each in a given color family. This is not strictly the case. There can be many variations in color and physical properties within color families, particularly with chemical lakes and toners, based on the dye content and the substrates used to make them. The noted differences will be in color strength, shade, depth of masstone, transparency, ease of dispersion, and oil absorption. The chemist must be aware of these differences in order to choose the correct additive to meet the color and physical requirements of the product being formulated.

Color by Shade: Organics

Let's begin this review of color by shade with the organic colors, proceeding from violets through blacks, and then move to inorganic colors, again starting with violets and this time going through whites. As with other parts of this book, only those colorants most commonly used will be covered as primary examples in each given category.

Organic Violets

Table 4.1 lists the main organic violet colorants by type, shade, use, and positive and negative traits.

As can be seen, although **D&C Red No. 34 Ca Lake** is an external only color, it gives high color strength and therefore is very important as a shading

Table 4.1   Organic violets.

Color type	Color shade	Major uses	Positive	Negative
D&C Red No. 34 Ca Lake	Violet	Nail lacquer Blush	High color strength	US external only
D&C Red No. 33 Al Lake	Violet	Lipstick Blush	High color strength	Slight bleed in water
D&C Red No. 33 Zr Lake	Violet	Lipstick Blush	High color strength	Slight bleed in water Not permitted in European Union, China, or Japan
D&C Violet No. 2	Violet	Nail lacquer Blush	—	US external only High cost

component in red nail lacquer and as a main color in violet nail lacquer varieties.

The color shades of the two **D&C Red No. 33 Lakes** are very close to one another, so the importance of using universally approved color additives in today's market has diminished the use of the Zr Lake, which is not approved for use in either the European Union or Japan.

D&C Violet No. 2, while it is a vibrant color, has two negative attributes that limit its use: first, it is an external only colorant in the United States, and second, it has a relatively high cost compared to the other violets. This latter attribute is a notable one, as it is the difference, functionally, between Violet 2 and the equally vibrant D&C Red No. 34 Ca Lake, itself a US external only color, but more cost-efficient in usage.

Organic Reds

Red is enormously popular in the personal care and cosmetics global market. With so many lip, nail, and blush cosmetic products favoring red, this category of colorants is much larger than any of the others; therefore, it will be broken down into two subcategories: **blue shade reds** and **yellow shade reds**. Beginning with the blue shade reds, Table 4.2 lists the major color types, alongside their shade, use, and attribute information.

D&C Red No. 7 Ca Lake, the first listed colorant, is the most important of the organic reds used in decorative cosmetics, which, when coupled with its cost, make it cost effective. It is the main color used in lipstick, nail polish, and blush. The only negative about D&C Red No. 7 and the D&C Red No. 7 Lakes is a poor stability in water, precluding their use in most aqueous systems. As the products in which they are used are predominantly anhydrous, however, water stability is of little consequence. D&C Red No. 7 Ca Lakes are available in a wide variety of color shades, strength levels, masstones, and transparencies, depending on the amount and type of substrate used. As mentioned earlier, it is advisable that multiple additives be used when creating a product color. With that in mind, one may think of D&C Red No. 7 Ca Lake as the base additive around which other colorants rotate in order to create a red-colored product.

D&C Red No. 30 Lake is used primarily because of its unique "pink" shade. A secondary reason for its use is that it has excellent resistance properties with respect to light, heat, solvents, and water. It would be much more widely used, but its cost is very high compared to other organic reds.

Table 4.2   Organic reds – blue shade.

Color type	Color shade	Major uses	Positive	Negative
D&C Red No. 7 Ca Lake	Blue red	Lipstick Nail lacquer Blush	High color strength Good economics	Slight water solubility
D&C Red No. 30 Lake	Pink red	Lipstick Blush	Unique shade Excellent stability	High cost
D&C Red No. 27	Blue red	Lipstick	Staining properties	Light stability
D&C Red No. 28	Blue red	Lipstick	Staining properties	Light stability
D&C Red No. 27 Al Lake	Blue red	Lipstick	Unique shade	General stability
D&C Red No. 28 Al Lake	Blue red	Lipstick	Unique shade	General stability
Carmine	Blue red	Lipstick Eye makeup	Natural Only true red approved for eye area use in United States	High cost Light stability

D&C Red No. 27 is the free acid of tetrabromo tetrachlorofluorescein (tetrabromo = four bromine atoms), while its cousin, **D&C Red No. 28**, is the sodium salt of the same molecule. Both of these colors are alkali soluble dyes and will stain the skin. This staining property is both these colors' greatest asset and a big obstacle to their wide usage. The main use for these dyes is in the formulation of long-wearing lipsticks, as they will stain the lips, remaining in the skin when the remainder of the lipstick film is removed through the daily routine of the wearer, giving the appearance that the lipstick wears longer. They are *not* recommended for use in most other types of decorative products, as the staining would be very unwelcome in a nail polish or blush! Novelty products have been made in which low percentages of fluorescein dyes are incorporated into a lip color or blush in order to make a claim of a "self-adjusting color" that the wearer can see develop on the skin.

When formulating for lipsticks, **D&C Red No. 27 Lake** and **D&C Red No. 28 Lake**'s unique color points, which are very close to each other, make it almost impossible in some cases to reach desired color shades without using one or the other of these colorants. A case in point is a fuchsia shade lipstick; in order to achieve the true fuchsia tone, one of these lakes must be used, because there is no other combination of US approved colors that can produce the shade. As these pigments are true chemical lakes, they will have a tendency to exhibit a slight bleed, which will result in a staining of the skin. While this is not a major issue in lipsticks, it can be problematic in other types of decorative products. Therefore, again, they find very little use outside of the lip area. D&C Red Nos. 27 and 28 are the bluest of the three colors in this family of fluorescein products. (The other two will be covered in the section on organic yellow shade reds and oranges.)

Carmine is unique among the majority of the natural colorants as it exhibits both a much brighter blue shade of red and a higher degree of intensity than any of the others in its class. There are some stability issues with carmine, as there are with other natural colors, but its main limitation to widespread use is its high cost. It is made by laking carminic acid on aluminum hydrate. Carminic acid is extracted from the female cochineal beetle, a species that is mostly indigenous to Peru, though it is also found to a lesser extent in the Canary Islands. Because of the limited supply and natural source, the price is high, and batch-to-batch variation can be significant. The main reason that carmine has continued to be used in color cosmetics is that it is the only true red pigment permitted for eye-area use in the United States. In eye makeup, when a bright, clean blue shade red is needed to match a desired shade, this colorant is the only way of producing it. Were the cochineal beetle more pervasive in its habitation and dissemination, carmine might find much wider use, or at the least, wider consideration among color chemists. Finally, carmine is also used where a natural colorant is needed or desired.

Now, on to the yellow shade reds. Table 4.3 lists the most commonly utilized examples.

D&C Red No. **6 Ba Lake**, the first listed, is the second most important of the reds, be they blue or yellow shade, used in decorative cosmetics. The majority of its use is in lipsticks, nail lacquer, and blush. It has the same chemical structure as D&C Red No. 7 Ca Lake, the difference being that the blue shade color is the calcium salt, while this is the barium salt. In addition to chemical structure, D&C Red No. 6 Ba Lake also shares with its blue shade counterpart the combination of a bright, "clean" color and a high degree of intensity, making it very good money value and thus suitable for common use. The major issue is that it is not permitted for use in cosmetics sold in Japan. (Want to know what is a good substitute, and presently the most often used, in that country? Keep reading.)

In **D&C Red No. 21** and **D&C Red No. 22**, which are the yellow shade versions of D&C Red No. 27 and D&C Red No. 28, one has two color

Table 4.3 Organic reds – yellow shade.

Color type	Color shade	Major uses	Positive	Negative
D&C Red No. 6 Ba Lake	Yellow red	Lipstick Nail polish Blush	High color strength and good economics	Water stability
D&C Red No. 21	Red	Lipstick	Staining properties	Light stability
D&C Red No. 22	Red	Lipstick	Staining properties	Light stability
D&C Red No. 21 Al Lake	Yellow red	Lipstick	Unique shade	General stability
D&C Red No. 22 Al Lake	Yellow red	Lipstick	Unique shade	General stability
FD&C Red No. 40 Al Lake	Yellow red	Lipstick Eye makeup Blush	Only synthetic organic red approved for eye-area use in the United States	Not permitted in Japan

additives with excellent, if limited, applicability and, consequently, fairly specific product applications. The former is the free acid of tetrabromofluorescein and the latter is the sodium salt. They are both alkali-soluble dyes, and like their bluer counterparts are used almost exclusively for making longer-wearing lipsticks that satisfy their advertised function by staining the skin. They are, however, not normally added to any other decorative products because of their staining nature.

D&C Red No. 21 Al Lake and **D&C Red No. 22 Al Lake** are pigments of D&C Red No. 21 and D&C Red No. 22. For the same reasons explained for the blue shade red colors D&C Red No. 27 Al Lake and D&C Red No. 28 Al Lake, these two pigments are found almost exclusively in lipsticks, where the tendency they have to bleed and stain is not a major factor in their use and, in point of fact, is seen as a benefit. The use of these colorants in other decorative products is, of course, severely limited by their bleed tendency. However, they exhibit very clean shades, which makes them quite useful in delicately colored lipsticks and lip glosses.

Finally, **FD&C Red No. 40 Al Lake** is a brick red that found very little usage in decorative cosmetics until 1994, because the color and economic properties of D&C Red No. 6 Ba Lake and D&C Red No. 7 Ca Lake were far superior. In the late 1980s, the Cosmetics, Toiletry and Fragrance Association, now known as the Personal Care Products Council, filed a petition – which the United States Food and Drug Administration (FDA) eventually approved, based on safety data submitted – calling for four certifiable colors to be permitted for use in eye makeup (Federal Register 1994). Along with FD&C Red No. 40 and its lakes, these were FD&C Blue No. 1 and its lakes, FD&C Yellow No. 5 and its lakes, and D&C Green No. 5. With this major change in regulations – the first time certifiable colors were approved for eye makeup in 60 years – the usage of FD&C Red No. 40 Al Lake increased, due to its superior batch-to-batch color consistency and lower cost over carmine. However, carmine is still superior in terms of color point, which somewhat blunts the usage of the FD&C Red No. 40 Al Lake. Another factor that limits its usage is that it is not permitted for use in cosmetic products sold in Japan. In that market, D&C Red No. 7 Lake is the color of choice for bright red eye-area cosmetics.

Organic Oranges

Table 4.4 outlines the major organic orange additives.

D&C Red No. 36 is on the border between red and orange, so it can be used as a shading color on either side to make reds yellower or to make yellows redder. Unlike the major red toners, it has good water stability, so it can be used in those rare instances where a decorative product is an aqueous system. It is not recommended for use in nail lacquer because of its tendency to bleed in solvents.

D&C Red No. 6, like D&C Red No. 36, is right on the cusp of being a red, so it can be used in the same way in terms of shading reds and yellows.

Table 4.4   Organic oranges.

Color type	Color shade	Major uses	Positive	Negative
D&C Red No. 36	Red orange	Lipstick Blush	Stable in aqueous systems	Solvent solubility
D&C Red No. 6	Red orange	Lipstick Nail lacquer Blush	Universal colorant	Weak color strength Water solubility
FD&C Yellow No. 6 Al Lake	Yellow orange	Lipstick Nail lacquer Blush	Bright shade	Slight bleed in aqueous systems
D&C Orange No. 5	Yellow red	Lipstick	Staining properties	Light stability
D&C Orange No. 5 Al Lake	Yellow red	Lipstick	Unique shade	General stability

However, the main use for it is as a substitute for D&C Red No. 6 Ba Lake (as previously mentioned), which is not permitted in Japan. While it is yellower and weaker than the barium lake, D&C Red No. 6 is nonetheless the best alternative there is. Like the other toners, it is not water-stable.

In **FD&C Yellow No. 6 Al Lake**, a true orange is found, with usage in formulas where an orange is needed to shade reds or yellows or where orange is the desired final color of the cosmetic product.

D&C Orange No. 5 completes the palette of fluorescein dyes that are used to make long-wearing lipsticks. It is a dibromofluorescein, meaning it contains two bromine atoms, and it is the yellowest of the three chemistries in the family of colors. Like the others, it is not used in other decorative products because of its staining properties.

Finally, **D&C Orange No. 5 Al Lake**, like its bluer fluorescein cousins, is only used as a shading component in lipsticks. As a true chemical lake, it would bleed and cause unwanted staining issues in any other type of color cosmetic.

Organic Yellows

There are limited number of accessible, usable organic yellows, as attested to by the rather truncated Table 4.5. However, it does not necessarily follow that these are lesser colorants, or that their usage is something less significant than their redder or yellower cousins.

FD&C Yellow No. 5 Al Lake is the workhorse organic yellow, the major yellow for shading in lipstick, nail polish, and eye makeup. It is another one of the four synthetic organic colors approved for use in eye makeup by the FDA in 1994 (Federal Register 1994). When it is combined with FD&C Blue

Table 4.5   Organic yellows.

Color type	Color shade	Major uses	Positive	Negative
FD&C Yellow No. 5 Al Lake	Red yellow	Lipstick Nail lacquer Blush Eye makeup	Only synthetic organic yellow approved for eye-area use in the United States	Slight bleed in aqueous systems
D&C Yellow No. 10 Al Lake	Green yellow	Lipstick Nail lacquer	Unique shade	High cost

No. 1 Al Lake, a very bright, clean, and intense green is produced, which can be used in the area of the eye. This is one of the situations where the rule commonly understood by chemists against mixing extremes of color doesn't apply, and, in fact, disobeying it bears quite wonderful results.

The second of our organic yellows, **D&C Yellow No. 10 Al Lake**, is a clean green shade yellow. Its use in decorative cosmetics is sadly limited because it is very expensive. In most cases, a formulator can use the FD&C Yellow No. 5 Al Lake as a shading color, which is much less expensive than D&C Yellow No. 10 Al Lake and yields satisfactory results. It is only used where unique shade characteristics are needed. A second reason for D&C Yellow No. 10 Al Lake's limited use is that it is not permitted for use in cosmetics products to be sold in the European Union (see Chapter 2).

Organic Blues

Admittedly a limited category, as can be seen from Table 4.6, the sole colorant under discussion here nonetheless bears significant properties that make its use key for color chemists.

FD&C Blue No. 1 Al Lake is the only practical choice for coloring decorative cosmetics among the organic blues, because the other notable one, D&C Blue No. 4, is approved for external use only in the United States, the small size of which market says everything about its limitation of usage. FD&C Blue No. 2 is not permitted in any cosmetic product sold in the United States. FD&C Blue No. 1 Al Lake is a very bright, strong color additive, and is another of the four synthetic organic colors approved by the FDA for use in eye makeup in 1994. As already mentioned, a very bright, intense green can be obtained when this colorant is mixed with FD&C Yellow No. 5 Al Lake.

Table 4.6   Organic blues.

Color type	Color shade	Major uses	Positive	Negative
FD&C Blue No. 1 Al Lake	Green blue	Lipstick Nail lacquer Blush Eye makeup	Only synthetic organic blue approved for eye-area use in the United States	Slight bleed in aqueous systems

Organic Blacks

Organic black is another smallish category, as can be seen from Table 4.7, but again a significant one, if for no other reason than the plethora of decorative cosmetics that aim at various subtle shades of black.

D&C Black No. 2 is a carbon black produced by the oil furnace process (FDA n.d.) and is a very intense jet black, exhibiting about 20 times more strength than black iron oxide – a useful property in achieving very black mascaras and eye liners. The major drawback to this colorant is that it is very difficult to handle because it dusts significantly. To avoid this problem, a number of pigment suppliers make D&C Black No. 2 available in dispersion form.

D&C Black No. 3 is also a carbon black, but it is produced from burning animal bones, as opposed to the petroleum-based D&C Black No. 2. It exhibits the same color properties as D&C Black No. 2, but is limited in use because of its animal origin. It is only included here due to the fact that formulators can use it as a natural ingredient.

Color by Shade: Inorganics

Now, before heading into color by application, we move on to the second overview of colors by shade, this one of inorganic colors. Once again, we'll begin with the violets and move forward from there.

Inorganic Violets

As described in Table 4.8, **Manganese violet** is available in both blue shade and red shade versions. Unlike several of the other inorganic pigments reviewed in this chapter, manganese violet is permitted for use in mucous membrane contact cosmetics in the United States, so it is used in lipsticks to

Table 4.7 Organic blacks.

Color type	Color shade	Major uses	Positive	Negative
D&C Black No. 2	Jet black	Eye makeup	Color strength and jetness	Difficult to handle
D&C Black No. 3	Jet black	Eye makeup	Color strength and jetness	Difficult to handle Animal source

Table 4.8 Inorganic violets.

Color type	Color shade	Major uses	Positive	Negative
Manganese violet	Blue violet	Eye makeup Lipstick	Low cost Approved for eye-area use in the United States	Lack of intensity Not stable in aqueous systems
	Red violet	Eye makeup Lipstick	Low cost Approved for eye-area use in the United States	Lack of intensity Not stable in aqueous systems
Ultramarine violet	Violet	Eye makeup Color correcting face makeup	Low cost Approved for eye-area use in the United States	Lack of intensity External only in the United States Not stable in acid systems

tone down bright organic colors, particularly in shades designed for darker skin. Its most common application is in eye shadow, where it produces quite pleasing violet shades without being too overwhelming in terms of intensity. Manganese violet will convert from the violet Mn^{+7} to the brown Mn^{+2} on contact with water, particularly at alkaline pH values, so it cannot be used in aqueous systems. Even the addition of small amounts of water to pressed powders to aid compaction during filling is not recommended.

Ultramarine violet is a weaker alternative to manganese violet, and is less used for several reasons. First, it is a US external only color, so it cannot be used in lipstick. Furthermore, batch-to-batch color consistency is poor, color intensity is low, and it is not stable in acid systems.

Inorganic Reds

The color additives in this category, as listed in Table 4.9, all share similar usage characteristics that are not precisely ideal. Still, they are, in their way, useful colorants.

Red iron oxide is the principal inorganic red used in a variety of decorative cosmetic products. Available shades range from yellow to maroon. Like manganese violet, it is used in lipsticks and nail polish to tone down the bright organic colors in products formulated for darker skin types and is a common colorant for eye makeup and lip color, particularly in earth-tone

Table 4.9   Inorganic reds.e

Color type	Color shade	Major uses	Positive	Negative
Red iron oxide	Yellow red	Lipstick Nail polish Eye makeup Foundation	Low cost Stability	Dirty shades
	Blue red	Lipstick Nail polish Eye makeup Foundation	Low cost Stability	Dirty shades
	Maroon	Lipstick Nail polish Eye makeup Foundation	Low cost Stability	Dirty shades
Ultramarine pink	Pink	Eye makeup	Low cost	Weak shade External only in the United States Not stable in acid systems

shades. The major use for red iron oxide, when blended with black iron oxide and yellow iron oxide, is to produce skin color shades in all types of foundation. The relative weakness of the oxides is mitigated by their preferred mixture with other colorants.

Ultramarine pink exhibits a nice pink shade, but like its cousin ultramarine violet, it is not widely used because of weak color strength, regulatory restrictions, and acid stability issues. Its predominant use is in eye shadow and highlighters, where the pink shade has the most value.

Inorganic Oranges

Orange iron oxide, a blend of red and yellow iron oxides, is the only choice in this category, as shown in Table 4.10. The only advantage of this color, aside from its relatively low cost, is the convenience of not having to blend the neat ones to arrive at an orange shade.

Inorganic Yellows

Yellow iron oxide is again the only choice in this category, as shown in Table 4.11. It is available in both green and red shades, and while often used

Table 4.10   Inorganic oranges.

Color type	Color shade	Major uses	Positive	Negative
Orange iron oxide	Orange	Lipstick Eye makeup	Low cost Stability	Weak, dirty shade

Table 4.11   Inorganic yellows.

Color type	Color shade	Major uses	Positive	Negative
Yellow iron oxide	Green yellow	Lipstick Eye makeup Foundation	Low cost Stability	Weak, dirty shades
	Green yellow	Lipstick Eye makeup Foundation	Low cost Stability	Weak, dirty shades

Table 4.12   Inorganic blues.

Color type	Color shade	Major uses	Positive	Negative
Ultramarine blue	Green blue	Eye makeup Foundation	Low cost	Moderate tinting strength External only in the United States Not stable in acid systems
Ferric ammonium ferrocyanide	Red blue	Lipstick Eye makeup Foundation	High tinting strength Deep masstone	Difficult to disperse External only in the United States Not stable in alkaline systems

in eye shadow to produce earth tones and in lipstick to dampen the vibrant color of organic colorants, it finds its major use when blended with red and black iron oxides to formulate skin tones for liquid makeup and other foundation products.

Inorganic Blues

Table 4.12 outlines the major inorganic blue color additives.

Ultramarine blue is the most common inorganic blue used in eye makeup. While not as strong as ferric blue, it is preferred by formulators due to its ease of dispersion.

Ferric ammonium ferrocyanide, commonly called **iron blue**, is used in eye makeup, but to a lesser extent than ultramarine blue due to dispersibility issues. Masstone is extremely dark, so iron blue can be used in mascara to make black iron oxide appear "blacker" up to levels of approximately 10% before the blue tone starts to show. Due to the rigorous methods of dispersion employed in the manufacture of nail polish, the dispersion issues of iron blue are marginalized, making it the blue shading component of choice in these lacquers.

Inorganic Greens

Table 4.13 outlines the major inorganic green additives.

Chromium oxide green is a yellow shade green that exhibits an olive drab or "dirty olive" look. Its predominant use is in eye shadow, typically blended with iron oxides to produce earth-tone shade products.

Chromium hydroxide green is a blue shade green and, for an inorganic color, displays a clean, bright look. It is most commonly used when formulating green or turquoise eye shadows, eye liners, or mascaras, where a clean, blue shade has value.

Inorganic Browns

Brown iron oxides, as listed in Table 4.14, are blends of the neat red, yellow, and black colors and are available in shades ranging from tan through light brown to chocolate brown. They are used for their ease of formulation, as they require only slight adjustments with the neat iron oxides to produce a wide range of shades, particularly the skin tones required for liquid makeup.

Table 4.13   Inorganic greens.

Color type	Color shade	Major uses	Positive	Negative
Chromium oxide green	Yellow green	Eye makeup Color-correcting face makeup	Low cost	External only in the United States Dirty olive green
Chromium hydroxide green	Blue green	Eye makeup	Clean shade	External only in the United States High cost High oil absorption

Table 4.14   Inorganic browns.

Color type	Color shade	Major uses	Positive	Negative
Brown iron oxide	Brown	Eye makeup Foundation	Low cost	Blend
	Tan	Eye makeup Foundation	Low cost	Blend

Inorganic Blacks

Table 4.15 breaks down the two versions of **black iron oxide**, really the only choice for inorganic black. It is available in a standard (black) shade that is used in conjunction with red and yellow to formulate a variety of skintone foundation shades and earthtone shades for eye shadow. A jet, bluer black version is also available, used almost exclusively to produce mascara and eye liners, where the deep jet black has value. Interestingly, in a tint, the jet black iron oxides are actually weaker than the standard browner black products.

Inorganic Whites

Titanium dioxide, as indicated in Table 4.16, is the white pigment used in virtually all types of decorative cosmetic products to provide coverage, dampen the intensity of other colors, and produce pastel shades, particularly in lipsticks and nail lacquer. While at first glance it appears that white is white, there are actually two different crystal forms of titanium dioxide, anatase and rutile, both of which are used in color cosmetics. Due to tighter packing of the atoms in the crystal structure, rutile titanium dioxide is harder, more chemically stable, more dense, and has higher bulk density compared with anatase.

Cosmetic grade titanium dioxide is defined as not less than 99% by US and EU regulations. Generally, such material exhibits good stability in cosmetic applications.

Table 4.15   Inorganic blacks.

Color type	Color shade	Major uses	Positive	Negative
Black iron oxide	Brown black	Eye makeup Foundation	Low cost	Weak tinting strength
	Jet (blue) black	Eye makeup	Low cost	Weak tinting strength

Table 4.16 Inorganic whites.

Color type	Color shade	Major uses	Positive	Negative
Titanium dioxide – anatase	Blue white	Lipstick Eye makeup Nail lacquer Foundation	Low cost	Lower chemical stability than rutile
Titanium dioxide – rutile	Yellow white	Lipstick Eye makeup Nail lacquer Foundation	Low cost	More difficult to suspend and disperse than anatase

In addition, titanium dioxide can have three different dispersion characteristics, based on the permitted <1% trace constituents present. The first is a pure material that is oil- and water-dispersible. The second is an oil-dispersible product that contains a small amount of aluminum oxide, which retards "floating" in lipsticks. The third grade disperses easily in water due to a high residual salt content, but has the disadvantage that floatation is sometimes observed in water-based foundation formulations.

Color by Product Application

Now that the various major use colorants have been reviewed, this chapter will move to the second phase of discussion, a review of color usage by type of cosmetic product. Covering lipstick, blush, eye shadow, mascara, liquid makeup, nail lacquer, and cosmetic soaps, this review, like the previous sections, will address the major colors applicable to each product segment.

Lipstick

With regard to the use of color additives, lipsticks are the most complicated decorative cosmetic products to formulate. There are a large number of colorants with good stability in lipsticks, giving the cosmetic chemist tremendous latitude in creating finished product shades. It is therefore not unusual to see 20 or 30 shades in a single product line – hence, the aforementioned difficulty. The color additives used in lipsticks are divided into several categories:

Main Colors: As the name indicates, these colorants normally make up the largest percentage of the color additive portion of the formula. They are the main colorants because they are bright and exhibit very good color intensity, making them good value for money. The colors in this category are:

D&C Red No. 6 Ba Lake

D&C Red No. 7 Ca Lake

D&C Red No. 6

D&C Red No. 6 Ba Lake is used as the main colorant in yellow red formulations, while D&C Red No. 7 Ca Lake is used in blue shade reds. D&C Red No. 6 is used as the main color in yellow shade reds for the Japanese and worldwide markets, as the barium lake is not permitted in Japan.

Shading Colors – Yellow: These colorants are used in conjunction with the main colors to create various shades in the yellow color space:

FD&C Yellow No. 5 Al Lake

FD&C Yellow No. 6 Al Lake

D&C Red No. 36

Either FD&C Yellow No. 6 Al Lake or D&C Red No. 36 can also be used as the main colorant in orange lipsticks.

Shading Colors – Blue: These colorants are used in conjunction with the main colors to create various shades in the blue color space:

D&C Red No. 33 Al Lake

D&C Red No. 27 Al Lake

D&C Red No. 28 Al Lake

D&C Red No. 33 Al Lake can also be used as the main color in violet lipsticks, while D&C Red No. 27 Al Lake (or D&C Red No. 28 Lake for Japan) is the color of choice for fuchsia shades.

Specialty Colors – Yellow: These color additives are used with the main and yellow-shading colors to create a wide range of shades in a given product line. They are weaker, and in some cases more expensive than the former two categories:

D&C Red No. 21 Al Lake

D&C Red No. 22 Al Lake

D&C Yellow No. 10 Al Lake

D&C Orange No. 5 Al Lake

Specialty Colors – Blue: These color additives are used with the main colors and blue-shading colors to create a wide range of shades within a given product line. Like their yellow counterparts, they tend to be weak and expensive:

FD&C Blue No. 1 Al Lake

D&C Red No. 30 Lake

Staining Colors: This is one of the few examples were dyes are used in decorative cosmetic products. They are used in conjunction with pigments to produce long-wearing lipsticks. Being alkali-soluble, they will stain the lips, giving the impression that the lipstick itself is longer-wearing:

D&C Orange No. 5

D&C Red No. 21

D&C Red No. 27

D&C Orange No. 5 is the yellow shade of the three colorants, D&C Red No. 27 is the blue shade, and D&C Red No. 21 sits right in the middle.

Inorganic Colors: Though used in smaller percentages than the organics, inorganic colors are quite useful in lipsticks. The common ones are:

Titanium dioxide

Iron oxides

Manganese violet

Mascara/Eyeliner

Inorganic colors find the most common usage in this area, simply because the vast majority of mascaras and eyeliners produced are black. Another reason is the very limited number of synthetic organic colors that the FDA permits in any eye makeup.

Inorganic Colors:

Black iron oxide (jet black)

Brown iron oxide

Ferric ammonium ferrocyanide

Ultramarine blue

Chromium hydroxide green

Manganese violet

Organic Colors:

D&C Black No. 2

FD&C Yellow No. 5 Al Lake

FD&C Blue No. 1 Al Lake

Due to its extreme jetness, D&C Black No. 2 provides the opportunity for the formulator to grab the holy grail of mascara: a very black shade. Often, to keep costs down, black iron oxide and D&C Black No. 2 are used together. The combination of FD&C Yellow No. 5 Al Lake and FD&C Blue No. 1 Al Lake makes a very bright, clean green but will tend to bleed in water-based formulations.

Eye Shadow

For the same reasons as with mascara, most of the colorants used in eye shadows are inorganic, keeping in mind any stability restrictions in formulations containing water.

Inorganic Colors:
Ultramarine blue
Manganese violet
Yellow iron oxide
Red iron oxide
Black iron oxide
Brown iron oxide
Chromium oxide greens
Chromium hydroxide greens
Titanium dioxide
Ferric ammonium ferrocyanide
Organic Colors:
D&C Black No. 2
FD&C Yellow No. 5 Al Lake
FD&C Blue No. 1 Al Lake
FD&C Red No. 40 Al Lake
Carmine

Further, and again as with mascaras, organic colors produce much brighter, cleaner eye shadow shades than do inorganic ones. FD&C Red No. 40 Al Lake is not only brighter than Carmine, but also much less expensive. D&C Black No. 2 can be used in place of black iron oxide to produce dark shades of brown.

Blush Products

Based on the color nature and intended application of these types of products, the organic colors are the predominant ones in powder and anhydrous products. The inorganic colorants are used to "tone down" the brightness of the organic colors and to produce earth tone shades. .

Organic Colors:
D&C Red No. 7 Ca Lake
D&C Red No. 6 Ba Lake
D&C Red No. 6
D&C Red No. 34 Ca Lake
D&C Red No. 30 Lake
D&C Red No. 36

As with lipsticks, D&C Red No. 6 is used as a replacement for Red 6 Lake in products formulated for the Japanese market, where the latter is prohibited. D&C Red No. 30 Al Lake is commonly used in blush due to its very high resistance to color change caused by perspiration or processing conditions, making the cosmetic product longer wearing and the shade more reproducible.

Inorganic Colors:
Iron oxides
Manganese violet
Ultramarine blue
Titanium dioxide

Liquid Makeup

Since the shades of these products are basically skin tones, the color additives used are essentially all inorganic. The iron oxides produce the various shades, while titanium dioxide adjusts the coverage and shade darkness.

Inorganic Colors:
Red iron oxide
Yellow iron oxide
Black iron oxide
Titanium dioxide

Nail Lacquer

Organic and inorganic colors are used in nail lacquer, but the organics predominate because nail lacquer shades tend to be bright colors, necessitating the use of the bright organic pigments. The overall palette of organic colors is limited due to stability problems with the harsh aromatic solvents used in most nail polish systems.

Organic Colors:
D&C Red No. 7 Ca Lake
D&C Red No. 6 Ba Lake
D&C Red No. 6
D&C Red No. 34 Ca Lake
D&C Red No. 30 Lake
FD&C Yellow No. 5 Al Lake
FD&C Yellow No. 6 Al Lake

D&C Red No. 6 Ba Lake and D&C Red No. 7 Ca Lake are the main colors used, for the same reasons they are commonly used in lipsticks: bright shades and good money value D&C Red No. 6 is used for Japanese formulations.

Inorganic Colors:
Red iron oxide
Black iron oxide
Ferric ammonium ferrocyanide
Titanium dioxide

Soap

Both organic and inorganic colors are used in cosmetic soaps. Generally, the decorative soaps use the organics, while the common bath and deodorant soaps use the inorganics. This is because the decorative soaps tend to be bright colors, requiring the use of the organics, while many of the deodorant products have more subdued shades, which are better suited to the inorganic pigments.

Organic Colors:
D&C Red No. 30 Lake
D&C Red No. 36
FD&C Red No. 40 Al Lake
FD&C Yellow No. 6 Al Lake
FD&C Blue No. 1 Al Lake

Inorganic Colors:
Iron oxides
Chrome greens
Ultramarines
Titanium dioxide

The next chapter will address the economic issues facing color chemists, in terms of both the practical hands-on nature of their work and the larger issue of how markets and availability affect the degree to which specific colors can be and are used.

References

FDA. (n.d.). Code of Federal Regulations, Title 21 Part 74.2052. www.accessdata.fda
.gov/scripts/cdrh/cfdocs/cfcfr/CFRSearch.cfm?fr=74.2052 (accessed September 28, 2020).
Federal Register. (1994). V59 N32 (Feb 16, 1994) pp. 7635.

Chapter 5
Color Selection – Economics

I f this book had been written 35 years ago, the overarching subject of color selection would have had only three elements: regulations, stability, and esthetics. The fourth element, concerning economics, would not have come into play because, at that time, there were far fewer cosmetic companies in the market than there are presently, meaning less competition and therefore little concern about the cost of ingredients and packaging. Furthermore, the vast majority of global cosmetic manufacturing was in the United States and Europe, which placed almost all of the manufacturers on a level economic playing field.

Today, there are many more cosmetic companies, both in the more traditional Western markets and the burgeoning Eastern and Asian ones, and there is also now a huge manufacturing base in Asia, particularly in China and Taiwan, resulting in a much more cost sensitive environment in which these companies must operate. So, in today's competitive marketplace, it is essential for a formulator to consider the most economical way to reach the color point he or she desires for each shade in a decorative cosmetic product line.

As an introduction to the subject of the economic implications of color selection, a review of the early 2020 pricing in the US marketplace is appropriate. Tables 5.1 and 5.2 show the approximate relative pricing for the commonly used organic and inorganic colors, respectively. Please keep this

Coloring the Cosmetic World: Using Pigments in Decorative Cosmetic Formulations,
Second Edition. Edwin B. Faulkner. Edited by Jane C. Hollenberg.
© 2021 John Wiley & Sons Ltd. Published 2021 by John Wiley & Sons Ltd.

Table 5.1   Comparative approximate bulk price of organic pigments (United States) as of January 2020.

Color	Price (USD/kg)	Color	Price (USD/kg)
Carmine	500.00	D&C Black No. 2	58.00
D&C Red No. 6	55.00	D&C Red No. 30 Al Lake	195.00
D&C Red No. 6 Ba Lake (21%)	42.00	D&C Red No. 33 Al Lake	90.00
D&C Red No. 6 Ba Lake (42%)	48.00	D&C Red No. 34 Ca Lake	63.00
D&C Red No. 7	65.00	D&C Red No. 36	51.00
D&C Red No. 7 Ca Lake (19%)	34.00	D&C Orange No. 5	117.00
D&C Red No. 7 Ca Lake (38%)	38.00	D&C Orange No. 5 Al Lake	56.00
D&C Red No. 21	160.00	D&C Yellow No. 10 Al Lake	111.00
D&C Red No. 21 Al Lake	72.00	FD&C Yellow No. 5 Al Lake (15%)	27.00
D&C Red No. 22 Al Lake	72.00	FD&C Yellow No, 5 Al Lake (24%)	28.00
D&C Red No. 27	196.00	FD&C Yellow No. 6 Al Lake	32.00
D&C Red No. 27 Al Lake	100.00	FD&C Red No. 40 Al Lake	45.00
D&C Red No. 28 Al Lake	100.00	FD&C Blue No. 1 Al Lake	43.00

Table 5.2   Comparative approximate bulk price of inorganic pigments (United States) as of January 2020.

Color	Price (USD/kg)	Color	Price (USD/kg)
Manganese violet	43.00	Black iron oxide (jet, blue-black)	14.50
Chromium oxide green	22.00	Black iron oxide (brown-black)	14.25
Chromium hydroxide green	148.00	Red iron oxide	14.00
Titanium dioxide	12.00	Yellow iron oxide	17.00
Ferric ammonium ferrocyanide	19.00	Brown iron oxide	17.50
Ultramarine blue	15.00	Orange iron oxide	17.00
Ultramarine violet	30.00		
Ultramarine pink	50.00		

information in mind, as these tables will be used as references throughout this chapter. It must be noted by the reader that these are approximate prices for use in understanding the price relationships among the different pigments. Prices in any particular market will vary based on the country of origin, transportation, currency exchange rates, and duties.

With these figures in mind, it is now possible to review the economics of color usage in decorative cosmetics.

Dye Content

The cost of manufacturing pigments is no different than the cost of manufacturing other chemical substances. It can be divided into two parts: **raw material costs** and **conversion costs**. Raw materials include the major intermediates used in the manufacturing process, the substrates necessary in laking, and any other minor chemical materials needed to set the process conditions and facilitate the chemical reactions. The intermediates account for the majority of raw material costs – often, up to 85% of the total. Conversion costs include – but are not strictly limited to – labor, utilities, equipment depreciation, supplies, and packaging materials.

As discussed in other chapters, organic colors are available in a very wide range of dye contents, starting as low as 10% for lakes and going as high as 98% in the case of primary colors. The conversion cost to make these colors is constant across the range of dye contents; in other words, the labor, utilities, supplies, packaging, and so on do not vary as the dye content changes. The dye is the most expensive component of the color additive, so the higher it is in a particular product, the more expensive that product will be. But – and it's a big but – the cost increase is not directly proportional to the increase in dye content, because the conversion cost remains constant as the dye content increases.

This phenomenon can be seen in Table 5.1 by looking at the D&C Red No. 7 Lakes, the D&C Red No. 6 Lakes, and the FD&C Yellow No. 5 Lakes, all of whose costs retain a relative level despite alterations in concentration. The lesson in all of this is that higher dye content lakes are much more cost-effective than lower dye content ones, as their increase in intensity is far greater than their higher cost per kilogram.

Effectiveness of Dispersion

Chapter 6 will delve directly into a discussion of dispersion, but as a preview to that discussion and a means of addressing the term here, dispersion is defined as **the process that converts a "raw" pigment into a usable form, providing the best color and money values**. What this means is that the better the job done in the factory dispersing pigments, the more economical the finished cosmetic product will be. Again, this fact will be addressed directly and become clearer in Chapter 6.

Dry Pigments vs. Predispersed Pigments

Decorative cosmetic companies generally buy dry pigments and disperse them in their factories. However, pigments are also marketed in dispersed forms in a wide range of vehicles, including low-viscosity synthetic waxes, high-melt-point waxes, castor oil, water, and others. Tables 5.3–5.5 show three types of dispersions, along with their respective relative prices.

As is the case with dry pigments, dispersion prices will vary from country to country based on currency exchange rates, transportation costs, and duty rates. The decision to use pigment dispersions versus dry color can be a complicated one, but it boils down to convenience, color consistency, equipment, and technology. By way of illustration, think about cooking a meal from scratch versus popping a frozen one into the microwave. Intuitively, it's obvious that the microwave meal is more expensive: you are paying for the convenience of not having to put the dish together yourself, and you also are paying for the name brand that has prepared it for you. However, if you do not have the time to prepare a meal from scratch, do not know how to cook well, want your meal to turn out the same every time, or do not have all the pots and pans you require to cook a particular dish, the premade frozen meal is the way to go. Judging by the amount of freezer space allocated to frozen microwavable dishes in most grocery stores, the vast majority of the public often turns to these dishes for one or more of these reasons.

Table 5.3   Castor oil dispersions.

Color	Price (USD/kg)	Color	Price (USD/kg)
D&C Red No. 7 Ca Lake (37% pigment)	32.00	Iron oxide black (50% pigment)	20.00
D&C Red No. 6 Ba Lake (33% pigment)	36.00	Iron oxide red (50% pigment)	21.00
D&C Red No. 6 Na Salt (35% pigment)	52.00	Iron oxide yellow (50% pigment)	22.00
FD&C Blue No. 1 Al Lake) (40% pigment)	39.00	Titanium dioxide (50% pigment)	17.50
FD&C Yellow No. 5 Al Lake) (40% pigment)	24.00		
FD&C Yellow No. 6 Al Lake) (45% pigment)	38.00		
D&C Red No. 33 Al Lake (40% pigment)	79.00		
D&C Red No. 28 Al Lake (45% pigment)	115.00		
D&C Red 21 (Dye at 50%)	198.00		

Table 5.4   Synthetic wax dispersions.

Color	Price (USD/kg)	Color	Price (USD/kg)
D&C Red No. 7 Ca Lake (37% pigment)	40.00	Iron oxide black (65% pigment)	32.00
D&C Red No. 6 Ba Lake (33% pigment)	42.00	Iron oxide red (65% pigment)	31.50
D&C Red No. 6 Na Salt (35% pigment)	48.00	Iron oxide yellow (50% pigment)	30.00
FD&C Blue No. 1 Al Lake) (40% pigment)	47.00	Titanium dioxide (50% pigment)	23.50
FD&C Yellow No. 5 Al Lake) (40% pigment)	38.00	Ultramarine blue (50% pigment)	39.00
D&C Black No. 2 (25% pigment)	67.00	Ferric ammonium ferrocyanide (27.5% pigment)	38.00

Table 5.5   Water dispersions.

Color	Price (USD/kg)	Color	Price (USD/kg)
D&C Red No. 30 Talc Lake (30% pigment)	100.00	Iron oxide black (38% pigment)	20.00
D&C Red No. 6 Ba Lake (30% pigment)	42.00	Iron oxide red (36% pigment)	19.50
FD&C Blue No. 1 Al Lake) (40% pigment)	25.00	Iron oxide yellow (30% pigment)	22.00
FD&C Yellow No. 5 Al Lake) (25% pigment)	18.00	Iron oxide brown (35% pigment)	22.50
D&C Yellow No. 10 Al Lake) (15% pigment)	39.00	Ultramarine blue (35% pigment)	18.50
D&C Black No. 2 (25% pigment)	61.00	Titanium dioxide (40% pigment)	19.00
D&C Red No. 33 Al Lake (15% pigment)	41.00		

A cosmetic company will often turn to dispersions, for the same reasons that a person will choose the microwave route. Making dispersions at a cosmetic company requires good dispersion equipment (pots and pans) and the technology to produce good dispersions (cooking know-how) that will achieve consistent color results (the meal turns out the same every time). Finally, in the rush to make a product on schedule, using dispersions saves a significant amount of manufacturing time (getting the meal on the table quickly).

The price of dispersions is dependent on several factors, including the type of pigment used, the percentage of that pigment, and the vehicle system employed. The cost to produce the dispersion and the manufacturer's desired profit margin must also be considered. Thus, when all the calculations are complete, a dry color will always work out to be cheaper than a premade dispersion, because the cost of the dispersion will include the producer's conversion costs and profit margin. However, just like the microwave meal, the convenience and consistency of the dispersion bring tremendous value to the table.

The following are a few "compare and contrast" examples of how cost considerations directly impact a chemist's choice, or available palette, when formulating.

Black Iron Oxide vs. D&C Black No. 2

As reviewed in Chapter 4, D&C Black No. 2 is several times more intense and "blacker" than black iron oxide, but unfortunately its cost ($50.25/kg) compared to the cost of the oxide ($9.75/kg) exceeds the greater degree of strength by a fairly large margin. This is not to say that it is not used, because there are times when its superior jetness trumps its proportionally higher cost in mascara and eyeliner. As a way of having the best of both worlds – jetness and an economical formula – a mixture of the two blacks is often employed.

Jet Black Iron Oxide vs. Brown Black Iron Oxide

While the cost difference between jet black iron oxide and the browner shade black iron oxide is only $0.50/kg, this seemingly small difference can amount to significant expense if the two colors are used inappropriately at significant volume. The jet black is designed for use in mascara and eyeliner, where the "holy grail" is the blackest black possible; therefore, the slightly higher cost is warranted. However, if the jet black is used to blend with red and yellow oxides in earth tone eye shadows or liquid makeup, its jetness value is lost and the $0.50/kg extra cost is wasted, as brown black is not merely "good enough," but preferable. To exacerbate this issue, the jet black is the weaker of the two, meaning that more of it needs to be used in place of the brown black to achieve the same color point, so the actual cost difference is higher than $0.50/kg.

Blended Iron Oxides vs. Neat Iron Oxides

The principles established earlier regarding dry color versus dispersion apply to this comparison as well. Iron oxide blends are prepared by proprietary processes that yield the ultimate in an intimate mixture that will not streak when a finished cosmetic product is applied to the skin. This processing cost, along with the manufacturer's profit, is built into the cost of these blends, making them more expensive than the neat colors (see Table 5.2).

Like dispersions, blends cut down on both formulating and processing times as each shade of finished product, particularly liquid makeups, can be developed and manufactured without further shade matching by using preblended iron oxides. This convenience does, however, come with a price, so a conscious decision must be made when and when not to use them.

Carmine vs. FD&C Red No. 40 Al Lake

Referring back to Chapter 4 once again, the reader will remember that Carmine is a much better bright red than FD&C Red No. 40 Al Lake because of its bluer tone. However, as can readily be seen in Table 5.1, Carmine is significantly more expensive, so its use is relegated to red shades where its superior color properties warrant the higher cost.

D&C Red No. 30 Lake vs. Other Red Lakes

Chapters 3 and 4 discussed the excellent stability properties and unique color shade of D&C Red No. 30. A further glance at Table 5.1 quickly reveals the cavernous difference between its price and that all of the other red lakes. Therefore, it is almost exclusively used where its unique properties have value. The certification data in Table 2.6 further bear this out.

D&C Yellow No. 10 Al Lake vs. FD&C Yellow No. 5 Al Lake

D&C Yellow No. 10 Al Lake is a very clean green shade yellow, while FD&C Yellow No. 5 Al Lake is a less clean red shade yellow. There are very few decorative cosmetics that come in yellow, meaning that these two pigments are used most of the time as shading components. The cost of D&C Yellow No. 10 Al Lake is four times higher than that of FD&C Yellow No. 5 Al Lake. Therefore, unless the formulator absolutely needs the distinct color properties of the former, he or she should always "grab" the container of the latter from the shelf when starting work on a new product that requires a yellow pigment.

Chromium Hydroxide Green vs. Chromium Oxide Green

Chromium hydroxide green is a clean blue shade colorant, while chromium oxide green is a dirtier yellow shade one, often described as "olive drab." In addition to the stark differences between these shades, the hydrated (hydroxide) one costs almost seven times more than the anhydrous (oxide) one. With this fact established, it is clear that the formulator should choose the chromium oxide when a green is wanted for shading iron oxides in earth-tone eye shadows. The use of the hydrated green is reserved for eye shadows and eyeliners where its color properties are needed.

Package Sizes

Most color companies will offer different size packages for each product they sell, two common ones being 5 and 25 kg, with the latter the more common. The 5 kg package offers smaller consumers of pigments the opportunity to buy amounts that better match their consumption and product output. However, with this convenience comes a stiffer price, as the majority of pigment producers add a premium for the smaller package, ranging from $5.00 to $10.00 per kg. As with any business involving a product, managing inventory stock-keeping units (SKUs) is a difficult task; the more SKUs in a line, the more opportunity there is for not having the desired package size when a customer orders. In addition, multiple SKUs drive up the overall amount of stock kept in inventory, which means more working capital required to run the business. For these reasons, SKUs for pigments are kept to a minimum by the manufacturers. The reason there is such a high premium on smaller packages is that the pigment producers create the smaller packages by repacking stock from the larger ones. This is done to expedite packaging time in the factory, as it takes considerably more time to "pack out" small containers in production. As with many things in this "big box," Costco world, buying in bulk has its benefits.

With the generally more expensive organic colors, this premium does not add a significant percentage to the per kilo cost of the product, so in most cases it makes sense for the smaller consumer to buy the smaller pack sizes. However, when the premium is applied to the inorganic colors, particularly titanium dioxide and the iron oxides, the price for a pigment can increase by 50% or more, as shown in Table 5.2.

Chapter 6
Pigment Dispersion

There have been many excellent articles written on pigment dispersion across various publications over the years, the vast majority of which have been very theoretical in nature, using mathematical equations accompanied by graphs and charts to reinforce a given author's particular points of examination or contention. Unfortunately, they have also tended to be generic in nature with respect to the end use applications for which the dispersions are intended (Vernardakis 1985, 2001; Ulrich et al. 1994; Paterson 2004).

The present chapter will not repeat the mathematics and graphs, but instead concentrates on the practical aspects of making good quality dispersions of pigment, focusing on the techniques and equipment commonly used in the decorative cosmetic industry. In nearly all cases, images of the equipment will accompany the discussion, in order to give you, the reader, a method to visualize the physical process itself.

The Importance of Dispersion

Dispersion is the most important process for incorporating pigments into decorative cosmetics. Though this importance has been alluded to earlier in the book, those allusions were brief, and before we can (finally!) begin a study of the dispersion process, it is necessary to define exactly what dispersion is – to secure a working definition of it that will see us through the review to follow.

Coloring the Cosmetic World: Using Pigments in Decorative Cosmetic Formulations,
Second Edition. Edwin B. Faulkner. Edited by Jane C. Hollenberg.
© 2021 John Wiley & Sons Ltd. Published 2021 by John Wiley & Sons Ltd.

But rather than one, there are two definitions that will be presented here. The first is a descriptive definition that gets to the heart of what dispersion is all about and the second addresses the technical aspects of the process itself. A pigment, when received by a cosmetic company, is in a highly aggregated, agglomerated state that is of very little color or money value.

So, addressing this issue, the first definition of dispersion is as follows: **Dispersion is the process that converts a "raw" pigment into a usable form, providing the best color and money values.**

The second, and more technical, definition is: **Dispersion is the process of wetting, separating, and distributing particles in a vehicle.**

Figures 6.1 and 6.2 illustrate very clearly what dispersion does and why it adds tremendous value to a pigment. The pair of images on the left of each display are masstone, meaning that they are prepared with two components – pigment and castor oil. The pair on the right are tints – prepared by diluting the masstone with zinc oxide. The masstone contains two individual components – the one on the left is made simply by mixing the pigments with castor oil using a spatula, while the one on the right is made by properly dispersing the color into the oil. It is very easy to see visually the impact dispersion makes on the color, as the masstones of the mixed preparations are very light and lacking in opacity when compared to the dispersed ones. Likewise, the color strength of the tints prepared by simple mixing is extremely weak compared to that of the properly dispersed ones. Gloss is another important property of lipsticks, lip glosses, and

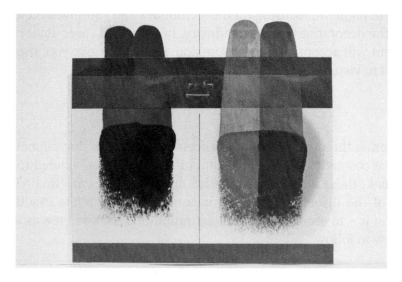

Figure 6.1 D&C Red No. 7 Ca Lake dispersion.
Source: Courtesy of Sun Chemical Corp.

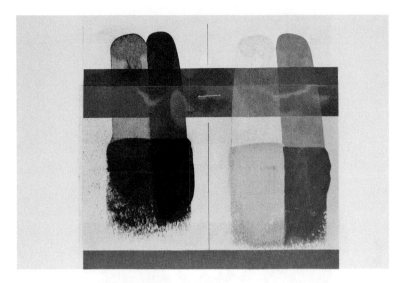

Figure 6.2   Ferric ammonium ferrocyanide dispersion.
Source: Courtesy of Sun Chemical Corp.

nail polishes that is affected by dispersion quality. The ability to provide shine on the skin is an important property of such cosmetics in terms of consumer appeal. Improperly dispersed pigments result in more diffuse reflection, which yields a dull finish. In addition, liquid cosmetic products must exhibit acceptable viscosity and flow properties, commonly called rheological attributes, in order for these products to be easily removed from their containers and applied smoothly on to the skin. Improperly dispersed pigments contain large particles that interfere with the rheology of the product, making it difficult to apply and causing it to rub off easily, forcing the consumer to reapply more frequently.

Pigment Particle States

Before embarking on a discussion of the dispersion process, a review of the various particle states in which pigments exist is in order, so that the reader can understand the fundamental changes that pigments must undergo in order to effectively be dispersed. Those states are as follows:

Primary Particles: These are net, discreet particles that exist only immediately after a pigment is formed during syntheses. They are submicron, meaning that they are <1 μm in size. The balance of the pigment production process, including the pressure of filtration and the heat cementing

Figure 6.3   Graphic representation of primary pigment particles.

Figure 6.4   Graphic representation of typical pigment aggregates.

properties of drying, force these primary particles together into much larger conglomerations called aggregates and agglomerations. Figure 6.3 provides a graphic representation of primary pigment particles.

Aggregates: These are pigment particles that form when the flat surfaces of primary particles come together and are cemented tightly together by pressure and heat. They are relatively small, tightly packed, and very difficult to break apart. Figures 6.4 and 6.5 are respectively a graphic representation and a photomicrograph that clearly shows the densely packed pigment aggregates.

Agglomerates: These are pigment particles that are formed in the production process when primary particles and aggregates come together at the points rather than at their crystal surfaces. Agglomerates are much larger than aggregates and are not as tightly bound, making them easier to break apart. Figures 6.6 and 6.7 are respectively, again, a graphic representation and a photomicrograph of some typical pigment agglomerates.

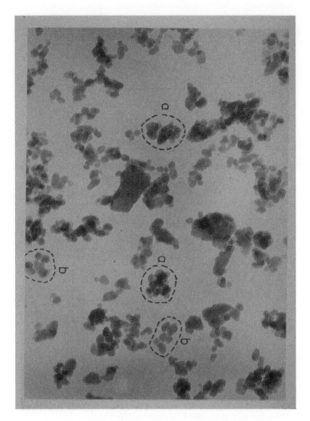

Figure 6.5 Aggregates in a photomicrograph.
Source: Courtesy of Sun Chemical Corp.

Figure 6.6 Graphic representation of typical
pigment agglomerates.

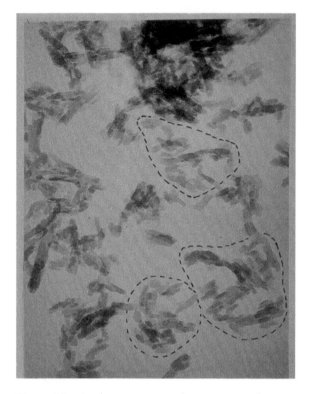

Figure 6.7   Agglomerates in a photomicrograph.
Source: Courtesy of Sun Chemical Corp.

The Dispersion Process

Pigment dispersion is a three-step process consisting of:

- wetting;
- separation/distribution; and
- stabilization.

The first step, **wetting**, is the process of replacing the air that is trapped in the spaces formed by the aggregation and agglomeration of the pigment particles with a vehicle. The spaces between the particles can be visualized in the representations in Figures 6.4 and 6.6. The vehicle softens the bonds within the particles, which facilitates their separation in the second step, separation/distribution. Vehicles are typically materials that are commonly used in the formulation for other purposes and which themselves have very good pigment wetting properties. The choice of vehicle depends largely on the type of cosmetic product into which the dispersion will be incorporated. Typical vehicles used in liquid or paste cosmetics include castor oil, silicones, esters, natural oils, natural and synthetic waxes, nitrocellulose, hydrocarbons, and water. For dry products, talc formerly was used almost exclusively, but concerns over safety have led to the use of alternative filler vehicles. There is no direct substitute for a given talc, but small particle fillers such as sericites, kaolin, and fluorophlogopite (synthetic mica) can be used, either singly or in combination. Dry dispersions are also made for finished cosmetic products such as press and loose powders, where talc or sericite is predominately used as the vehicle.

The variables that play a part in the effectiveness of wetting are the **pigment-to-vehicle ratio, surface tension**, and **time**. Oil absorption of pigments varies among the myriad chemistries used in decorative cosmetics. Typical values for the common pigment types can be found in Table 6.1, where the method used to determine them is **ASTM D 281-24**, which calls for the use of linseed oil as the vehicle. Please note that the values for castor oil will be higher, but the relative differences among the pigments will remain the same. It must also be noted that these are typical values; actual ones for specific products from different manufacturers will vary based on their individual characteristics, particularly among the organic pigments, which are available in a wide variety of dye contents and substrate types.

It is critical to have the correct **pigment-to-vehicle ratio** to ensure that there is enough to fill all the spaces between the particles, but not so much

Table 6.1   Oil absorption of common pigments (ASTM D 281-24).

FDA name	Linseed oil/ dry pigment (gm/gm)	FDA name	Linseed oil/ dry pigment (gm/gm)
Chromium oxide green	0.212	Ferric ammonium ferrocyanide	0.629
Russet iron oxide	0.224	D&C Red 6 Ba Lake	0.636
Red iron oxide	0.298	D&C Red 7 Ca Lake	0.708
Orange iron oxide	0.301	D&C Red 7 Ca Lake	0.719
Brown iron oxide	0.367	D&C Red 6	0.860
Titanium dioxide	0.400	D&C Orange 5	0.939
Black iron oxide	0.409	D&C Red 28 Al Lake	0.946
Orange iron oxide	0.443	D&C Red 21 Al Lake	0.967
Ultramarine blue	0.470	D&C Red 33 Al Lake	0.980
Manganese violet	0.499	D&C Red 36	1.060
Orange iron oxide	0.500	D&C Yellow 6 Al Lake	1.070
Black iron oxide	0.501	FD&C Yellow 5 Al Lake	1.290
Yellow iron oxide	0.575	D&C Red 27 Al Lake	1.323
Chromium hydroxide green	0.582	D&C Orange 5 Al Lake	1.431

that the viscosity of the dispersion is too low to obtain good separation in the second step of the process. This means that pigments with high oil absorption values will be wetted at higher oil-to-pigment ratios, while those with lower oil absorptions will be wetted at lower oil-to-pigment ratios.

The second variable in the wetting process is **surface tension**, which is defined as the ability of a liquid to resist the separation of its molecules, The lower the surface tension, the easier it will be for a given vehicle to fill the spaces among the pigment particles, so the use of surfactants, which serve to lower surface tension, is recommended in systems where they are compatible. Surfactants that improve pigment wetting are referred to as **wetting agents**.

Finally, because this step does not happen instantaneously, adequate **time** must be provided to allow the spaces among the pigment particles to be filled by the vehicle, followed by a softening of the particles. Again, this ensures a more effective separation/distribution in the next step of the dispersion process.

The equipment utilized for wetting is, of course, going to be specific, functionally, to wet and dry dispersions, respectively. For the former, either a high-speed mixer equipped with a Cowles type blade (Figure 6.8) or a low-speed mixer equipped with a propeller (Figure 6.9) is commonly used.

Figure 6.8   High-speed mixer with Cowles blade.

Figure 6.9   Low-speed mixer with propeller blade.

If the low-speed mixer is chosen, the wetting time must be proportionally extended to the reduced rate of speed.

When wetting a dry dispersion, a 50/50 blend of the pigment and talc (or other filler) must first be mixed in a blender designed for dry powders, which will usually be of the ribbon (Figure 6.10), paddle (Figure 6.11), conical, or "V" variety. Such blenders are necessary because of the non-fluid nature of dry powders. As they do not mix without agitation, the use of tumbling action and agitators that cover all areas of the blenders is necessary to insure homogenous blending of the components.

Once the pigment is thoroughly wetted, it is time to move on to the second part of the dispersion process: **separation/distribution**. This step takes the wetted aggregates and agglomerates and separates them into their smaller constituents, meaning that the agglomerates are separated into smaller aggregates and primary particles, and the aggregates are separated into primary particles to the extent possible. These smaller particles are then distributed evenly throughout the mass of the vehicle.

Figure 6.10   Ribbon blender.
Source: Courtesy of Color Techniques, Inc.

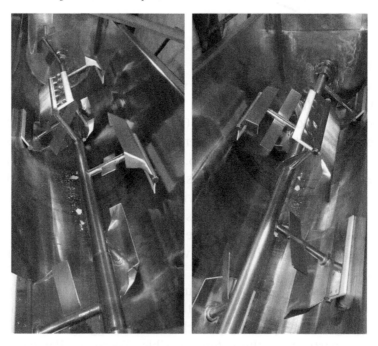

Figure 6.11   Paddle blender.

Figure 6.12   Hammer mill.
Source: Courtesy of Color Techniques,
Inc.

Separation/distribution, not unlike wetting, can be broken down into more than one form of dispersion – in this case, into liquid and powder dispersion. We'll start with the latter.

Following dry blending, powders are processed through a dry mill, typically the hammer and screen type (Figure 6.12), to finish wetting and achieve separation and uniform pigment particle distribution. The speed of the mill and the size of the screen selected are determined by the hardness and crystalline structure of the pigment being dispersed. Softer materials, such as inorganic pigments and low dye content organic pigments, are milled through slotted screens in the 0.035″ range. Hard materials, typified by the high dye content organic colors, are milled through holed screens in the 1/8–1/4″ range. Several passes through the mill may be required to achieve the desired particle size in the pigment dispersion. Visual evaluation is a simple, direct means of determining the adequacy of a dry pigment dispersion. A wide spatula is an effective means of preparing drawdowns by hand. A sample of the dispersion is placed on heavyweight textured white paper and is drawn down with firm pressure using the edge of the spatula. Undispersed color shows up as darker streaks in the drawdown.

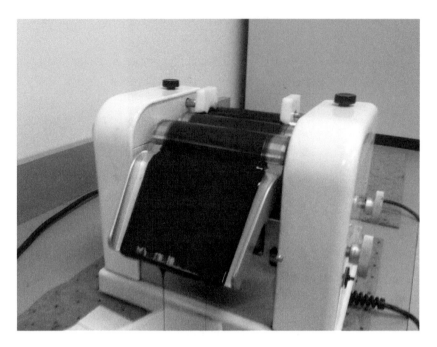

Figure 6.13   Three roll mill.

There are two types of mills used for making liquid dispersions: shear mills and media mills. The shear mills are typified by the roller types, most having either two or three rolls. Two roll mills are used to make dispersions of very high-viscosity materials such as nitrocellulose and high-melt-point waxes like polyethylene. Three roll mills are used to make dispersions of somewhat lower-viscosity materials such as castor oil. Examples are seen in Figures 6.13 and 6.14.

The rolls on the shear mills rotate at different speeds, which creates the shear on the wetted pigment, effectively separating the larger particles into smaller ones. In a typical three roll mill operation, the wetted pigment is introduced into a space called the nip, between the second and third rolls counting backward from the front of the mill. The difference in the roll speeds will cause the material being milled to transfer to the nip between the second and first rolls, creating more shear and therefore more particle separation. Finally, as the material transfers to the first roll, it will be scraped off by a very sharp piece of metal called a doctor blade into a container by means of a chute, called an apron. This process is repeated as many times as is necessary to produce the desired pigment particle size. Each time wet-ted pigment is processed through the mill, it is called a pass; the number of passes required will depend on the dispersibility of the pigment type. These

Figure 6.14   Two roll mill.
Source: Courtesy of Sun Chemical Corp.

mills are by far the most effective ones for wet dispersions, but they only work for higher-viscosity vehicles, because dispersions with low viscosities will flow between the rolls, resulting in no effective separation and a big mess to clean up.

In order to determine when the separation process is completed, it is necessary to measure the size of the pigment particles. While there are some very sophisticated analytical instruments available to perform this task, it is much easier and more efficient in a factory setting to use a grind gauge, such as a Hegman gauge (Figure 6.15).

A Hegman gauge is a rectangular block, constructed of stainless steel, into which two parallel channels have been cut. The channels shown in Figure 6.15 gradually increase from a depth of $5\,\mu m$ at the bottom of the gauge to one of $100\,\mu m$ at the top. The two non-converging channels ensure consistency of the readings. To use the gauge, a small amount of the dispersion is placed into the top of each channel and then smoothly – in one fluid motion – dragged down to the bottom by means of a blade held perpendicular to the flat surface of the gauge. The operator, or technician, will see scratches on the surface of the channels made by the particles where they exceed the depth of the channels; the closer the scratches to the top of the gauge, the larger the particles. A good dispersion should have particle sizes in the range of $12–13\,\mu m$, which is represented by graduation marks

Figure 6.15   Hegman gauge.

of approximately 7–8 on the gauge. Some manufacturers specify a lower particle size range for pigment dispersions intended for incorporation into lipsticks due to the tactile sensitivity of the lips. The measurement process is repeated after each pass on the shear mill until the specified size is met.

The second type of mill employed to make liquid dispersions is the media mill, sometimes called a bead mill because of the shape of the medium. As the name implies, a media mill is a closed vessel with media inside, into which the wetted pigment is introduced under pressure, causing an interaction with the aggregates and agglomerates, breaking them down into the smaller particles typical of a good dispersion. The pigment slurry can be passed through the mill into a receiving vessel or, more commonly, recirculated as many times as is necessary to produce the desired degree of dispersion.

Pigment grinding media are typically round or cylindrical, composed of ceramic, stainless steel, or glass, and come in sizes starting at approximately 1 mm. Size and material of construction are determined by the type of pigment and the vehicle system used to process the dispersion.

The method for determining completed separation when using media mills is different from that employed for shear mills. Because the viscosity of media milled material is much lower, preparation of proper drawdowns on a Hegman gauge is not possible. Instead, a color evaluation is performed at the tint level, comparing the color strength of the batch to an established standard. This process is explained thoroughly in Chapter 7. When the two are equal, it is assumed that the particle separation is complete. While not as effective or efficient as shear mills, media mills work much better for low viscosity vehicles.

Irrespective of the type of mill being employed, there are variables that control the effectiveness of the separation/distribution process: the amount of energy employed, the time utilized for the process, the viscosity of the wetted pigment, and the avoidance of polychromatic dispersions. First, the more energy exerted on the aggregates and agglomerates, the more effective their separation. This energy is determined by the type of mill employed. Shear mills apply more energy to the wetted pigment than do media mills, which explains why the former are more effective in the creation of dispersions. Second, the slower the speed at which wetted pigment is passed through a mill, the more opportunity there is to perform particle separation. There is often a misconception on the part of a given company's manufacturing department that speeding up the mill shortens the processing time. This is untrue, and unfortunately makes it necessary to increase the number of passes the pigment slurry must make through the mill, actually prolonging the overall processing time. Third, the viscosity of the wetted pigment plays the same role in the separation process as it does in the wetting step, as discussed earlier in the section on wetting.

Finally, the last variable in the separation/distribution process is the age-old controversy of making monochromatic versus polychromatic dispersions. Monochromatic dispersions are made using only one pigment, while polychromatic dispersions are made with more than one pigment, and often several. On the surface, it seems logical that if all the pigments to be used in one shade of a given finished product are wetted and dispersed together, a significant amount of processing time can be saved, as opposed to making a number of dispersions with individual pigments. The fallacy in this thinking is that due to the differences in wettability and dispersibility of the various pigments, the maximum color value cannot be obtained using the polychromatic approach. A visual explanation for this phenomenon can be found in Figures 6.16 and 6.17. Both graphs plot color strength (K/S values) on the vertical axis against the number of mill passes on the horizontal one (note that each plot's results are realized on a three-roll mill). In Figure 6.16, a D&C Red No. 7 Ca Lake must be processed between four and five times before it reaches its maximum color strength, while the

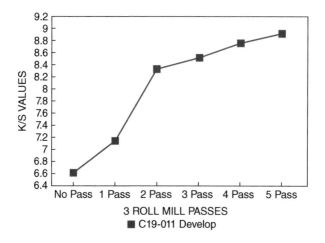

Figure 6.16   Strength vs. time: D&C Red No. 7 Ca Lake.

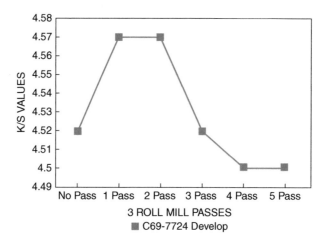

Figure 6.17   Strength vs. time: FD&C Yellow No. 5 Al Lake.

FD&C Yellow No. 5 Al Lake in Figure 6.17 does so after only one or two passes and actually loses strength during subsequent passes. Therefore, if a polychromatic dispersion containing these two pigments is made and stopped at two passes, the D&C Red No. 7 Ca Lake will not reach its maximum color potential, nor its most effective money-to-use ratio. By the same token, if the same polychromatic dispersion is made using FD&C Yellow No. 5 Al Lake at four passes, value, both esthetic and monetary, will be lost.

The final step in the dispersion process is **stabilization**. Once dispersed, pigment particles will tend to move toward one another, causing them to reagglomerate, resulting in a loss of strength and a change in hue. One result of this phenomenon, called flocculation, is a loss of color strength and a change in hue. The reason for the color change is that flocculated color contains larger particles with reduced surface area, causing a reduction in light reflection. A pigment's tendency to flocculate is largely dependent on particle size, particle surface charge, and the viscosity of the finished product in which it is dispersed. This problem – particularly undesirable with liquid foundation makeup – can manifest itself when a cosmetic is displayed on a store shelf. The bottle tone of products containing flocculated pigments is not representative of the skin tone that develops during rub out on the skin, a distinct negative for consumers who choose what shade to purchase based on observed bottle tone.

In order to avoid flocculation, it is necessary to stabilize products with one or more materials designed for this purpose. In aqueous systems, the common stabilizers are clays and organic materials, such as gums, synthetic polymers, proteins, cellulose, and plant polysaccharides. For non-aqueous systems, metal soaps, waxes, silicone crosspolymers, dextrin palmitate, and inorganic stabilizers including natural clay minerals, organo-modified clays, and synthetic silicas are most often employed.

Ed Faulkner's Note: Though by no means meaning to denigrate the importance of any aspect of this book, I nonetheless consider this chapter to contain the most vital information. Dispersion is the cornerstone of formulating for decorative cosmetics. If you, the reader, fully understand the principles laid out and practice the techniques described herein, the number of color issues you will encounter will be reduced significantly.

References

Paterson, J. (2004). The art of mixing: high speed dispersers. *Paint and Coatings Industry.*

Ulrich, D., Schak, J., and Kritzer, R. (1994). Optimizing the dispersion process with pre-milling. *Paint and Coatings Industry.*

Vernardakis, T.G. (1985). Improving dispersion of organic pigments. *Modern Paint and Coatings* 75 (9): 32–42.

Vernardakis, T.G. (2001). Pigment dispersion. In: *Coatings Technology Handbook* (eds. D. Satas and A.A. Tracton). New York: Marcel Decker.

Chapter 7

Color Measurement and Pigment Testing

Pigment Testing

This chapter will delve into the qualitative and quantitative ways to measure the differences among batches of pigments and will deal with the principles of and methods used to test pigments for use in decorative cosmetics. The adage "like peas in a pod" – meaning to bear similar if not identical characteristics, behaviors, traits, or interests – is familiar to almost everyone. When it is applied to color additives, it is true only with regard to their chemical nature. The batch to batch chemical fingerprint of any color additive will be identical; spectrophotographic curves will show essentially no variation in the chemical structure. However, due to differences in particle size, particle size distribution, and degree of agglomeration, the color values of a particular color additive can vary distinctly between batches. Such variations lead to color consistency problems in the manufacture of cosmetic products, resulting in excess color adjustments along with attendant delays in the production process.

The proper application of some basic principles to the day-to-day evaluation of color additives prior to their use in cosmetic production will serve to minimize many of these consistency problems. Therefore, this chapter will address the quality control (QC) of incoming batches of color additives, as opposed to the evaluation of colors for specific formulations.

Coloring the Cosmetic World: Using Pigments in Decorative Cosmetic Formulations, Second Edition. Edwin B. Faulkner. Edited by Jane C. Hollenberg.
© 2021 John Wiley & Sons Ltd. Published 2021 by John Wiley & Sons Ltd.

The six basic principles of interest to us are as follows:

Selection and Validation of a Test Method: In order to eliminate apparent differences in results between testing laboratories, it is very important that they use a common test methodology. This principle is especially important when trying to compare results between the color additive manufacturer's test facility and one belonging to the cosmetic producer who is evaluating the color manufacturer's pigment prior to purchase. As with any type of experimental work, the fewer the variables involved in the process, the better the chances of success. Further, it is important to ensure that the test method is reproducible. This means that the same batch, when tested several times, will always produce the same results.

Establishment of a Standard: In order to objectively evaluate the acceptability of batches of any given color additive, it is essential that they be compared to the same standard every time. Since the batches will vary slightly from one to another, the practice of adopting the last batch used in production would cause havoc in the evaluation process, akin to shooting at a moving target. The standard established for evaluation of incoming receipts should be selected in consultation with their manufacturer. Color additive manufacturers have master standards against which they evaluate their productions, as well as acceptable tolerances on either side that allow them to approve or reject batches in a consistent manner. These tolerances, also called specifications, can be subjective (if the batches are judged by the human eye) or objective (if they are judged by a color computer).

The necessity of consulting the manufacturer of a color concerning standard selection is illustrated in Figures 7.1 and 7.2. In Figure 7.1, a color manufacturer would approve both batch A and batch B for shipment. The two batches are both within the established tolerances, though they are on opposite sides of the standard. Figure 7.2 illustrates what happens if a cosmetic producer arbitrarily selects batch A as the standard and then receives batch B as a shipment. Even though both of these batches are within commercial tolerance of the color manufacturer's standard, they are very far apart from each other. The cosmetic producer would perceive the difference as an inability on the part of the color manufacturer to produce additives consistently, when, in reality, the problem is one of standard selection.

Preparation of the Standard: In order to save time and lab effort, it appears on the surface that it would be best to prepare a large dispersion of the standard in the appropriate system and use it to test numerous batches of color additives as they are received. Unfortunately, this practice

Figure 7.1   Left: batch strength 95% of standard; Center: standard; Right: batch strength 105% of standard.
Source: Courtesy of Sun Chemical Corp.

will lead to problems and erroneous results in testing. The major reason for this is that the color additives will continue to "wet out" over time, as the vehicle will work its way into the spaces between the agglomerates, causing them to separate. The result is a greater surface area, which allows more light to be reflected, resulting in more color strength. The increase in strength of a standard dispersion of color over time will again provide the technician with a moving target at which to aim.

The other reason that the making of a large dispersion of a standard leads to problems is that it introduces new variables into the situation: the dispersing equipment and the technician. Each piece of dispersing equipment (e.g., a Hoover Muller) will disperse color additives to different degrees, giving different results in terms of color values. If the standard and the sample are made on two different pieces of equipment, one cannot be certain whether any color differences seen are a result of a real difference from the standard or only of differences in the dispersing equipment.

Figure 7.2   Left: batch strength 95% of standard; Right: batch strength 105% of standard.
Source: Courtesy of Sun Chemical Corp.

Likewise, using two different technicians to run the standard and sample
might also lead to erroneous results, due to subtle differences in the way
they perform the test procedure. Therefore, it is imperative to prepare the
standard and the sample fresh each time, using the same equipment and
the same technician. This will eliminate the possibility of color differences
coming from inconsistencies in these two sources.

Evaluation in the Appropriate System: Color additives will not perform
equally in different end use systems. For example, a high dye content D&C
Red No. 6 Ba Lake will yield one result when used in nail lacquer and
another when used in a blush. Colors of this nature are difficult to disperse, so they will appear weak and dull in powder applications because the
dispersion process normally employed is not very rigorous. On the other
hand, the nitrocellulose chipping process used in making dispersions for

nail polish is very effective, so the high dye content colorant will appear bright and strong. With this type of difference in mind, it is necessary to test color additives in the same system in which they will be used in production. If this principle is not observed, a batch that tests "OK" in the laboratory evaluation may perform poorly in the factory; conversely, a perfectly acceptable one may be rejected if the QC test does not really measure how it will work in production.

Evaluation at the Appropriate Use Level: As color additive performance is system-dependent, it is also affected by use level. Just as a color additive will exhibit different dispersion and color characteristics at disparate use levels, so an entirely different color profile will show itself when a pigment is used at 8–10% in a particular formulation than when it is used at 3–5%. Therefore, as with testing in a real system, it is also necessary to test color additives at the level at which they will be used in the actual production situation.

Maintenance of Constant Light Conditions: Have you ever observed how the color of your car looks different when it is illuminated by the sun than when it sits under yellowish sodium vapor lights in a parking lot at night? This phenomenon is called metamerism, which simply means that colored substances appear different in hue under different light conditions. Cosmetic color additives are no exception, so, in order to ensure consistency of color test results, it is necessary to always use the same source of illumination when observing test displays. There is no right or wrong, good or bad light source; rather, it is all about consistency. To maximize consistency, it is advisable to procure a light box with several different artificially generated types of light, including simulated daylight. By using such a device, it is possible to eliminate the vagaries of natural sunlight and the changes in spectral output of incandescent and fluorescent lightbulbs as they age.

Observance of these six principles will serve to minimize the possibility of test results varying among laboratories. In addition, following them closely will prevent many of the color problems commonly encountered during the cosmetic product manufacturing process.

Test Methodologies: Three Basic Methods

What follows are three basic test methods that can be employed to evaluate batches of colored pigments before they are used in production, as well as one (staining color evaluation) for the specialty dyes and one (dry color eval-uation) that is not recommended at all. The procedures are described here, while the actual test methods can be found in Appendix A.

Castor Oil Evaluation: This test is used to evaluate the general color prop-erties of a colorant, including depth of masstone, transparency, shade, and strength. It specifically measures a colorant's performance in oil systems, particularly lip products and emulsions where the colors are incorporated into the oil phase. The test involves the dispersion of the colorant sample and standard in castor oil, utilizing a muller – a machine that disperses small quantities of pigment into a vehicle (Figure 7.3). The grinding sur-face consists of two glass plates, between which the pigment and vehicle are sandwiched. The bottom plate is stationary, while the top one rotates. The action of the plates serves to disperse the pigment into the vehicle. Sub-sequently, side-by-side (touching) thick and thin film drawdowns of the two materials are prepared on specially designed white paper striped with a black line. Small amounts (puddles) of the two materials are placed on the upper portion of the paper and then drawn down using a wide-bladed knife, which is very similar to a common putty knife. Light pressure is applied to the knife during the first part of the downward stroke, and heavy pressure as the materials are passed over the black line. The thick film area over the white portion of the paper is used to evaluate the masstone of the colorant, while the thin film area over the black line is used to determine the degree of transparency. Together, the films on the paper are referred to as a "display", as example of which is shown in Figure 7.4.

The next step in the procedure is to dilute the castor oil dispersions made in the first part with a white pigment, commonly zinc oxide or titanium dioxide. Once again, a display is made by drawing down the two pastes, side by side, on the drawdown paper in both thick and thin films. The shade and strength of the test sample can then be evaluated, predominantly using the thick-film area of the display, as shown in Figure 7.5.

Talc Evaluation: This test is used generally to measure the dispersion properties of a pigment and specifically to evaluate the performance of pigments in powder systems, such as blush, eye shadow, and face pow-der. The procedure is to disperse both the sample and the standard (sepa-rately, of course) into talc using a standard kitchen-type blender. The make and model are not important, but the speed at which the samples are dis-persed and the time employed are critical to obtaining consistent results. The speed should be the highest setting on the machine that will not cre-ate a dust cloud in the laboratory; the time is normally 60–90 seconds. The

Figure 7.3   Hoover Muller (Source: Sun Chemical Corp.).

color and talc dispersions are then placed side by side (touching) on a piece of white paper, covered by a layer of tissue paper, and flattened with a wide-bladed knife. The tissue paper is removed and the color differences in terms of shade and dispensability are determined, as shown in Figure 7.6. **Nitrocellulose Evaluation:** This test is used to evaluate the performance of a colorant in nail enamel. The first step is to disperse the color and standard separately into nitrocellulose on a two-roll mill, incorporating a

Figure 7.4   Castor oil masstone display.
Source: Courtesy of Sun Chemical Corp.

plasticizer into the dispersion. For those not familiar with nitrocellulose, its main commercial use is in gunpowder and explosives, so extreme caution must be exercised: it is highly flammable and, under the right conditions, highly explosive. The safety instructions included in the test method must be followed meticulously.

The pigment dispersion will come off of the mill in a sheet, which is then broken into small pieces, commonly referred to as nitrocellulose or N/C chips. These chips are then cut into a solvent mixture containing acetates, alcohols, and additional nitrocellulose. The "cutting in" process is accomplished using a high-speed, high-shear mixer or a common paint shaker of the type you might see in the paint and priming section of your local hardware store. The resulting lacquers are displayed side by side on an 8×11.5 in. card, commonly called a Morest chart. This chart is white on the upper half and black on the lower. The display is made in a similar manner to the castor oil display, except that a specially designed bar is used

Figure 7.5 Castor oil tint display.
Source: Courtesy of Sun Chemical Corp.

Figure 7.6 Talc display.
Source: Courtesy of Sun Chemical Corp.

is ensure a standard thickness of the films throughout the entire process (Figure 7.7). This display is used to judge the depth of the colorant's masstone over the white area of the chart and its transparency over the black area (Figure 7.8).

Figure 7.7   Drawdown bar.
Source: Courtesy of Sun Chemical Corp.

| D & C Red 7 Ca Lake | FD & C Yellow 5 Al Lake | D & C Red 34 Ca Lake |

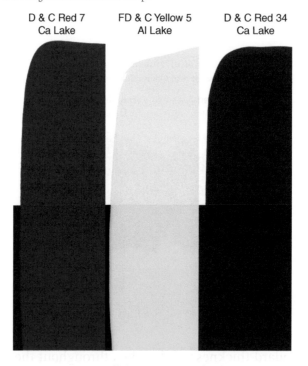

Figure 7.8   Nitrocellulose masstone display.
Source: Courtesy of Sun Chemical Corp.

As in the castor oil test, the prepared lacquers are diluted with a white pigment, normally zinc oxide. The dilution, called a tint, is made by means of a high-speed mixer or the paint shaker employed in the first part of the test. A drawdown of the diluted color sample next to the standard is prepared in the same manner as the masstone display by using the drawdown bar to make a uniform film thickness of the two samples. The tint is used to judge the shade and the strength of the test batch against the standard (Figure 7.8).

Staining Color Evaluation: This test is used to measure the staining power and shade of the halogenated fluorescein dyes used in the traditional manufacturing processes for long-wearing or permanent lipsticks. The list of dyes is short:

D&C Orange No. 5
D&C Red No. 21
D&C Red No. 22
D&C Red No. 27
D&C Red No. 28

| D & C Red 7 | FD & C Yellow 5 | D & C Red 34 |
| Ca Lake | Al Lake | Ca Lake |

Figure 7.9   Nitrocellulose tint display.
Source: Courtesy of Sun Chemical Corp.

As they are all pH sensitive, they are only weakly colored at a pH below their individual equivalence points, exhibiting their intense color and staining properties on the alkaline side. Therefore, the test method must involve incorporation of the dyes into a system with an alkaline pH.

The initial step is to disperse the test dye and the standard separately into an acrylic vehicle on a paint shaker using 1/8 in. steel shot as medium. Once dispersed, a let-down vehicle is added to each and mixed further on the shaker. The resulting dispersions are then displayed in the same manner as in the nitrocellulose test, utilizing the same type of Morest chart and rod to make the display films (Figure 7.10). This masstone display is used to make a visual assessment of the test batch against the standard.

Figure 7.10   Test sample vs. standard; masstone display.
Source: Courtesy of Sun Chemical Corp.

Figure 7.11   Test sample vs. standard; tint display.
Source: Courtesy of Sun Chemical Corp.

Like the liquid test procedures, the masstone dispersions are diluted with titanium dioxide white dispersion and mixed on the paint shaker. Once again, a side-by-side display of the two materials is made on the Morest chart using the prescribed rod. Visual assessments are made of the relative shade and strength differences between the test sample and the standard (Figure 7.11).

Dry Powder Evaluation: This test is not really a valid one and should therefore not be included in this chapter. However, it is discussed here because some cosmetic companies still use it as a method to evaluate batches of cosmetic colorants. This author wants to help all readers of this book understand that it has no value in color evaluation. The reason that it has no value is that it violates two of the principles outlined earlier. Before discussing which ones, it will be useful to describe the test itself.

The arrangement of the samples is similar to that of the talc evaluation, in that a row each of the full strength standard and the sample are laid side by side (touching) on a piece of white paper, covered by a layer of tissue paper, and flattened with a wide-bladed knife. The tissue paper is then removed and the differences between the two materials is observed.

Now, with a description of the test completed, the reader will be able to understand which of the principles are violated by using this test method. The first is **evaluation in the appropriate system**. There is no system here. The first three tests described in this section are specifically aimed at a batch's performance in a specific application system (e.g., lipstick, powder, nail polish). This test does not call for the colorant to be dispersed in a vehicle; it is simply a side-by-side comparison of dry powders. Any variation in agglomerate size between the samples will result in a different apparent color, even if, when properly dispersed, the color is a good match.

The second principle violated when using this test is **evaluation at the appropriate use level**. Absorption pigments are never used as neat 100% materials in a cosmetic product; there are always other ingredients present.

With these factors in mind, the author urges the reader to resist the temptation of comparing samples of pure color to evaluate how a batch of a colorant will perform under factory conditions. The only appropriate use of this "quickie" test is to identify the basic family of a colorant (e.g., is an unidentified drum of material in the factory a D&C Red No. 6 Lake or a D&C Red No. 7 Lake?). Comparison of two raw colors serves no other useful purpose.

Again, for ease of description of the foregoing methods, details of the test methods and materials used have been omitted. They can be found in Appendix A.

A final point to be noted in the area of display preparation for color testing involves placement of the sample and standard. Of course, there is strictly speaking no right or wrong approach to placement, but most companies will place the standard on the left and the sample on the right. The key here is to be consistent. No matter what relative positioning is chosen for the two materials, it must be employed each and every time an evaluation is done, across all parts of the organization. If this is not done consistently, a great deal of confusion will be generated within the company.

Color Measurement

All of the displays described thus far are basically designed for judging color differences qualitatively, using the human eye as the measurement device.

In this part of the chapter, a slightly different display preparation will be discussed. This is the one employed when using a color spectrophotometer/computer as the measurement device.

We described Alfred Munsell's early-twentieth-century theory of color in Chapter 1. In brief, Munsell postulated that any color has three attributes that make it unique, and therefore different from any other color. These attributes are hue, brightness, and intensity. He developed a crude set of color cards, incorporating these three aspects, which could be used to measure color differences utilizing the human eye as the measurement device. Based on the available technology of the time, this technique was as sophisticated as Munsell could make it. It was not until the 1970s, seven decades after the institution of the Munsell system, that the modern age of color measurement was born as scientists merged two technologies, spectrophotometry and computers, to make color measurement truly quantitative.

One of the most influential organizations in the development of today's color measurement systems has been the Commission Internationale de l'Eclairage (CIE). From the Yxy color space developed in 1931 to the L*a*b color space introduced in 1976, which is still in use today, quantitative color measurement has made the process of communicating color a modern science. The development of these two systems – and others – can be compared to the evolution of sound recording from vinyl disk to eight-track tape to cassette tape to compact disk and on to today's digital rendering.

The most common color space systems that have been developed in the last 80 years are as follows:

XYZ Tristimulus Values: This system is based on the theory that the human eye is composed of receptors for the three primary colors – red, green, and blue – and that all other colors are combinations of them. While state of the art in 1931, it has little use today, as it only pinpoints hue in its measurement.

CIELab: This system, pronounced "see lab" was the first to provide for measurement of brightness and intensity in addition to hue. The current version is CIEL*a*b*, the most widely used system used for color measurement. Because of its wide usage, a more detailed examination of its workings will be covered later in this chapter.

CIEL*C*h: This system is a cousin of CIEL*a*b*. The difference is that CIEL*C*h locates a color in color space using polar coordinates rather than rectangular ones.

CMC: The Society of Dyers and Colourists' Color Measurement Committee (CMC) designed this system as a mathematical equation to measure color difference based on CIEL*C*h color space.

FMC-2: Like CMC, this system is a color-difference equation rather than a color space. It was devised by Friele, MacAdam, and Chickering (FMC).

Hunter Lab: This system, developed by R.S. Hunter, was designed as an improvement over the original CIE Yxy color space. It is similar to the CIEL*a*b* color space and is still used in some industries, particularly in paint and coatings.

CIEL*a*b*, A Detailed View

As mentioned, CIEL*a*b* color space measures all three attributes of color: hue, brightness, and intensity. Figure 7.12 shows the three-dimensional CIEL*a*b* color space, where brightness, named the "L*" value, is represented on the vertical axis, hue is represented on the horizontal plane, and intensity (chroma, the "C*" value) increases from the center of the wheel to its outer edges.

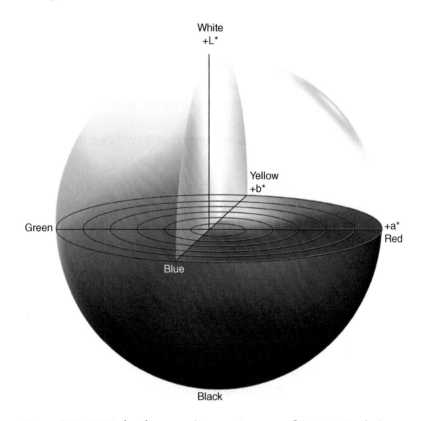

Figure 7.12   CIELab color space (Source: Courtesy of Konica Minolta).

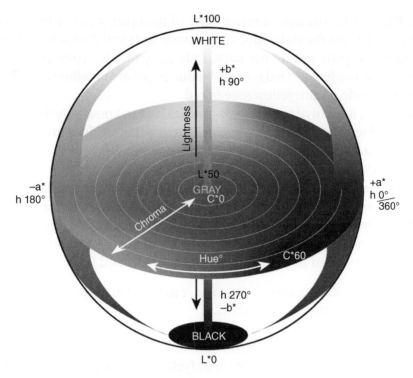

Figure 7.13   Cross-section of CIELab colorspace.
Source: © X-Rite Incorporated | www.xrite.com.

Figure 7.13 shows a cross-section of the three-dimensional CIEL*a*b* color space, making it easier to see the hue/chromaticity attributes, with the red/green axis being named the "a*" value and the blue/yellow one the "b*" value. The "L*" value is a number that increases from 0 (black) to 100 (white) according to the degree of lightness. The "a*" value can have a + (redder) or − (greener) sign. Similarly, the "b*" value can have a + (yellower) or − (bluer) sign.

To illustrate how CIEL*a*b* works quantitatively, refer back to the lipstick shown in Figure 1.7 and the corresponding light reflection spectrum shown in Figure 1.8. All wavelengths are absorbed, except for those corresponding to the color we perceive as the red shade of the lipstick. The spectrophotometric measurement of the wavelength of maximum reflection equaling 610 nm is converted by the color computer to the corresponding numeric CIEL*a*b* values:

$$L^* \ 43.31/a^* \ +47.63/b^* \ +14.12$$

In Figure 7.13, the approximate place of the lipstick in CIEL*a*b* color space would be on the right edge of the cross-sectional diagram, just above the midline, slightly toward the back, indicating that the color is near total saturation, strongly +a* (red), and b* value is positive (yellow), indicating an intense shade on the yellow side of red.

Of course, it is not of tremendous practical use to know only the absolute numeric values of a color; the use comes when these values are compared to a standard, enabling the user to judge the acceptability of the color for a particular application.

The beauty of the spectrophotometer and color computer is that the former measures the individual color values of the standard and sample separately, while the latter calculates the color differences, reports them to a technician, and stores them in memory for future use. The color difference parameters generated in the CIELab system are defined as follows:

DL*: This is the difference in brightness, expressed in terms of being lighter (+) or darker (−) than the standard.
Da*: This is the difference in the red/green axis, expressed in terms of being redder (+) or greener (−) than the standard.
Db*: This is the difference in the blue/yellow axis, expressed in terms of being bluer (−) or yellower (+) than the standard.
De*: This is the total difference in intensity (chroma), expressed in terms of being stronger (>1) or weaker (<1) than the standard.
DE*: This is the total color difference between the standard and the sample.

The color difference parameters used in the CIEL*C*h system are:

DL*: This is the difference in lightness from the standard.
DC*: This is the difference in chroma (intensity).
Dh*: This is the total difference in hue between the standard and the sample.
DE*: This is the total color difference between the standard and the sample.

Table 7.1 shows how a typical CIELAB color difference printout enumerating these parameters would look. The interpretation of the results is that the sample is 2.1 units darker, 0.52 units greener, 0.85 units bluer, and 1.2 units stronger than the standard. The total hue difference is 1.1 units and the total color difference is 2.32 units. Specifications can be established for all of these parameters and used to determine quantitatively the acceptability of a sample for use.

Table 7.1   CIELab and color difference printout.

	DE*	DL*	Da*	Db*	DC*	DH*
Differences	2.32	−2.1	−0.52	−0.85	1.2	1.1

Sample Preparation

The final topic to cover in color measurement is how samples are prepared for presentation to the aperture of the spectrophotometer. The tests described in the first part of this chapter produce side by side displays of the standard and sample that are used for visual evaluations of color. The displays prepared in the nitrocellulose acrylic tests are suitable for use in instrumental color measurement. However, those prepared in the castor oil and talc tests are not, so with these tests the resulting dispersions must be prepared for presentation to the spectrophotometer in a different manner. The test methods found in Appendix A contain instructions for the preparation of visual displays for instrumental analysis.

Chapter 8

Surface Treated Pigments

Origins of Surface Treated Pigments

The incorporation of surface treated pigments and substrates into decorative cosmetics dates back over four decades to the 1970s. The process was first introduced by Miyoshi Kasei in Japan (Faulkner and Zavadoski 1994), who gave formulators a whole new set of tools with which to enhance the esthetic and functional qualities of their formulations. The use of surface treated materials was slow to spread to the nations of the Western hemisphere, however, and it wasn't until the mid-1980s that they became popular in the United States and Europe.

In an effort to improve the durability of decorative cosmetics, silicone treated pigments and substrates were developed and introduced into Japanese formulations in 1976. The silicone treatment of these materials rendered them extremely hydrophobic, which improved the wear characteristics of the finished products, because they were much more resistant to the effects of perspiration. As an added benefit, the hydrophobic nature of the treated materials paved the way for the introduction of two-way pancake makeup into the Japanese market, which could be applied with either a wet sponge or a dry applicator through many applications, based on the consumer's preference at the time of use.

Though it offers a considerable improvement over conventional materials, there are some drawbacks to the silicone treatment. First, silicone treated pigments and substrates have a very low bulk density, which makes them somewhat difficult to press as powders. Second, silicone treated

Coloring the Cosmetic World: Using Pigments in Decorative Cosmetic Formulations,
Second Edition. Edwin B. Faulkner. Edited by Jane C. Hollenberg.
© 2021 John Wiley & Sons Ltd. Published 2021 by John Wiley & Sons Ltd.

materials tend to have a dry feeling, which can produce an unpleasant sensation when applied to normal to dry skin types. In an effort to address these deficiencies, a metal soap treatment was introduced to the Japanese market in 1977. Materials treated with metal soap have a higher bulk density than their silicone counterparts, and they are very useful in pressed powder formulations, where metal soaps are commonly used as binders. As an added benefit, metal soap treated pigments and substrates have a smooth, soft feeling when applied to the skin.

The next phase in the evolution of surface treated pigments came in 1979, when an amino acid system was introduced to the Japanese market, which brought improvements in skin feel over both the metal soap and silicone technologies. In addition, the amino acid chemistry helped to make cosmetic products – particularly liquid makeup – more compatible with skin chemistry, thereby producing longer wearing products.

As the tactile properties of cosmetic products grew in importance, a hydrogenated lecithin treatment was developed for the Japanese market in 1982. The original source of the lecithin was egg yolk, as this was more stable to oxidation than the material derived from soy. This treatment was – and, with further developments of the technology, still is – the epitome of a soft, smooth, and creamy feel during product application to the skin. Since 1982, the field of surface treatment of pigments and substrates has grown dramatically in terms of the number of companies supplying treatments to the cosmetics industry, the types of treatments available, and the number of cosmetic companies incorporating treated materials into their products.

Benefits of Surface Treatment of Pigments

The benefits of using surface treated pigments and substrates can be divided into two categories: those evident in the finished cosmetic product and readily discernable by the consumer, and those that accrue in the form of process improvements during manufacture of cosmetic products.

The benefits identifiable by the consumer are as follows:

Two-Way Cakes: The use of hydrophobic treated materials in pressed powders allows the user to apply the makeup at home with a wet sponge and then "touch it up" later with a dry applicator when no water is available; hence the name "two-way" cake. This combination product would not be possible with conventional raw materials, as the cake would become rock hard after drying.

Hydrophobicity: Hydrophobic treated pigments and substrates exhibit much better wear characteristics on the skin as they are significantly more resistant to perspiration than their untreated counterparts.

Skin Adhesion: The improved surface properties of most treated materials, coupled with their hydrophobicity, yield finished cosmetic products that exhibit better skin adhesion compared to those made with conventional materials. The result is less smudging and creasing, along with longer wear characteristics.

Tactile Properties: Most of the surface treatments impart a smooth, soft, creamy feel to the cosmetic products into which they are incorporated. This improvement in product feel, which is particularly noticeable in loose and pressed powder products, adds a new dimension to the value and attractiveness of the finished product as it is perceived by the consumer.

Suspension: Treating pigments with certain materials – mainly silicone and polyethylene – changes their surface characteristics, leading to a reduction of the age-old problem of the settling of titanium dioxide and iron oxides in nail lacquer.

Moisturization: Many of the surface treatments add an emollient, moisturizing quality to both pigments and substrates. This results in finished products that have a skin-care aspect as well as a decorative one.

Application/Spreadabilty: Ease of application, including smooth spreadability, is of increasing value to today's consumer. Finished products made with treated ingredients are noticeably easier to apply, spreading and blending more smoothly than traditional cosmetics.

The benefits that accrue in the manufacturing process include the following:

Dispersibility: In most formulations, surface treated pigments are easier to disperse than their untreated counterparts. This is because the treated surface prevents the pigment particles from re-agglomerating in processing and storage. Easier dispersion leads to shorter process time and lower energy costs.

Pressability: Most treated pigments and substrates are inherently easier to press. Surface treatment also facilitates the intimate mixing of the binders with other ingredients, improving pressability. This helps reduce processing time and enhances finished product quality.

Oil Absorption: Surface treating pigments reduces their ability to absorb oil. In anhydrous products, this phenomenon permits incorporation of higher pigment loads in a product of given viscosity, or allows the creation of a lower-viscosity product at a given pigment load. Also, by equalizing the oil absorption of pigments, fewer wax/oil ratio adjustments are required to produce a range of shades while maintaining physical properties – particularly useful in lipstick formulation.

Uniform Surface Properties: Surface treatment produces pigment and substrate particles with more uniform surface properties. This allows for better compatibility and more consistent performance among mixtures of pigments having widely differing surface chemistries, thereby reducing shade-to-shade differences in texture, stability, and performance. Better mixing of ingredients and thus shorter processing times are achieved in production.

Shrinkage: Because of their hydrophobic nature, surface treated pigments absorb less moisture, which is important in the manufacture of cosmetic pencils. This results in less moisture being released after manufacturing, reducing shrinkage of the core.

Commercial Surface Treatments

A wide variety of surface treatments have been introduced to the market over the last 40 years. Most are still available for use in the manufacture of decorative cosmetic products today. Some of the more common ones are:

Silicone (INCI name: Methicone): The first of the surface treatments, silicone still provides the highest degree of hydrophobicity of any treatment type available. It is also quite effective in keeping pigments in suspension in nail polish. Mineral oil can be added during the treatment process to reduce the skin drag that is sometimes experienced in pressed powder products with silicone alone. The major issue with Methicone treatment is that the cross-linking reaction generates hydrogen, which can continue to be evolved long after the treatment process is finished, leading to cracked cakes in pressed powder products and gas bubbles in liquid ones. Manufacturing procedures have been developed to drive the reaction between the Methicone and the substrate pigment to completion in order to avoid this post-treatment hydrogen evolution.

Silicone and Vegetable Hybrid: This treatment is a combination of Methicone and hydrogenated palm oil. It provides the strong hydrophobicity of the silicone along with a smooth, creamy feel and better adhesion to the skin. Materials treated with this "silky silicone" have better pressability than those treated with silicone alone.

Alkyl Silane: Triethoxycaprylylsilane is the most commonly used silane for cosmetic pigment and filler treatment, sometimes used as an alternative to Methicone or Dimethicone. Triethoxycaprylylsilane treated materials are more hydrophobic than those treated with Methicone and have better skin adhesion properties. The caprylyl chain on the molecule creates a surface that wets better in hydrocarbons and esters than do methylsilicone surfaces.

Metal Soap: This treatment has the best all-round economic value, as it provides good hydrophobicity along with a smooth, creamy feel when used in powder products. It also aids in the pressing of powders due to its ability to act as a binder.

Amino Acid: This treatment is especially well suited for liquid makeup formulations. In addition to its hydrophobic properties, which help improve skin adhesion and wear, the slight acidity resulting from the treatment renders pigments and substrates more compatible with the pH of human skin, thus extending wear. Materials treated with amino acid help preserve the delicate chemistry of the skin's surface by maintaining the sebum layer and suppressing bacterial growth.

Lipo Amino Acid: While similar to the standard amino acid treatment, this has superior properties in the area of skin adhesion, coverage, and skin feel.

Lecithin: Hydrogenated lecithin treated pigments and substrates set the standard of excellence for finished product feel. Lecithin exhibits the smoothest, softest, creamiest feel of all the available surface treatments. Hydrogenated lecithin, derived from egg yolk, avoids the odor and color problems associated with soy bean lecithin and contains a higher level of phospholipids. For formulators avoiding animal derived products, a vegetable lecithin treatment with enhanced stability is now available.

Lauroyl Lysine: This is similar to the amino acid treatment in most of its properties. It is principally employed for the treatment of substrates and pearlescent materials to improve skin feel and increase compressibility.

Esters: Isostearyl sebacate is used to treat pigments and substrates primarily for powder applications because it offers a wet-feeling payoff when applied to the skin while maintaining good properties of skin adhesion and wear.

Jojoba Esters: This is another popular treatment used as an alternative to "synthetic" compounds that imparts excellent hydrophobicity and skin adhesion properties.

Perfluorocarbon: Treatments based on polyfluoroalkylphosphate esters or fluoroalkyl silanes render pigments both hydrophobic and lipophobic. As a result, these materials are not affected by either oil or water in a formula, providing for improved wear. Other available polymers, such as perfluropolymethylisopropyl ether, impart similar hydrophobic/lipophobic properties to treated pigments. Due to current concern about the ubiquitous presence of perfluorooctanoic acid in the environment, use of shorter-chain perfluoro compounds is preferred by many to avoid the possibility of contamination by the compound.

Polyethylene: This treatment provides excellent skin adhesion while offering good miscibility into the oil phase of cosmetic emulsions. It has been used to advantage in nail polish products, where the modified surface prevents the settling of titanium dioxide and iron oxide pigments.

Silica: This treatment is used to prevent the darkening of pigments on the skin when they are exposed to water or oil.

Waxes: Carnauba wax is used as a treatment material because, in addition to good hydrophobicity, it provides a creamy feel and good adhesion when applied to the skin.

Metal Alkoxides: Mainly based on organic titanates, this unique type of hydrophobic treatment yields pigments and substrates that exhibit excellent skin feel when used in decorative cosmetic products. As with alkyl silanes, this treatment is useful in both dry and liquid formulations.

Polysaccharides: Galactoarabinan, a material naturally derived from trees, is the typical treatment in this category. It is hydrophilic, so the treated pigments wet well in aqueous systems, requiring very little high shear dispersion. The resulting particles are of such small size that they stay in suspension far better than untreated pigments.

Surface Treatment Mechanisms

There are four mechanisms by which pigments and substrates are surface treated:

Chemical: This method results in a chemical reaction between the treatment active and the surface of the pigment or substrate, forming a chemical bond between the two. Examples of treatments that work via this mechanism are Methicone and alkyl silane.

Mechanical: The mechanism for this type of treatment is a physical adsorption of the treatment active on to the surface of the pigment or substrate, without a chemical bond between the two. An example of a treatment that works via this mechanism is galactoarabinan.

Electrostatic: Electrostatic forces between the treatment active and the surface of the pigment or substrate bond the two materials together. Lauroyl lysine is a typical example.

Precipitation: A soluble form of the treatment active is precipitated on to the surface of the pigment or substrate, often by a di- or trivalent metal ion. Metal soap, lecithin, and amino acid treatments are examples of this category.

Performance Testing of Treatment Types

The following data show the relative differences among treatment types for several categories of performance-improvement testing. Some are based on quantitative performance oriented test results, while some are qualitative, based on panelist evaluations. All of these data, qualitative and quantitative, were designed, conducted, and reported by JCH Consulting, whose principal, Jane Hollenberg, has graciously allowed their use here for the discussion of color treatment.

Editor's Note: The following test results were part of work I performed some time ago to quantify the effects of surface treatment on pigment performance in actual products. Basic methodology, not requiring sophisticated instrumentation, was used, which formulating chemists can duplicate.

Compressibility: This factor is always an issue with pressed powder products. The ability of various surface treatments to impact it is shown in Figure 8.1. These data were compiled by surface treating all of the pigments and fillers used in a pressed powder foundation with each of various treatment actives and then compacting them into typical pressed powder pans. The pans were dropped from a prescribed height on to a hard surface repeatedly until the cake broke. The higher the numbers of drops a pan could withstand before its cake broke, the better its compressibility. The results demonstrate that the control, made with untreated materials, has the worst compressibility, while the silane treatment has the best.

Hydrophobicity: This attribute plays a role in adhesion to the skin and color fastness. As hydrophobicity increases, so does a product's ability to withstand the effects of perspiration. Figure 8.2 shows comparison data on several treatments with respect to their hydrophobic nature. The data were gathered by measuring the percentage of methanol required to reduce

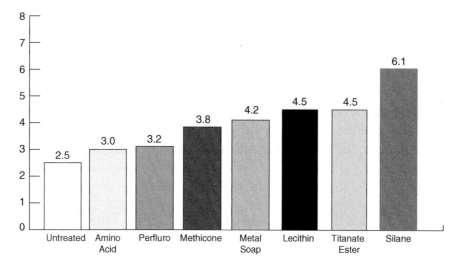

Figure 8.1 Compressibility.

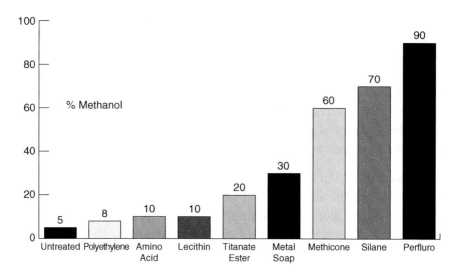

Figure 8.2 Hydrophobicity.

the hydrophobicity of a treated titanium dioxide pigment to zero, as evidenced by complete wetting of the powder when a set amount was added to the liquid. The higher the percentage of methanol required, the more hydrophobic the material. In this data set, the untreated control has the lowest hydrophobicity and the perfluoro treated material has the highest.

Oil Absorption: This property is one that must be managed carefully by the formulator to ensure that all shades within the shade range of a color

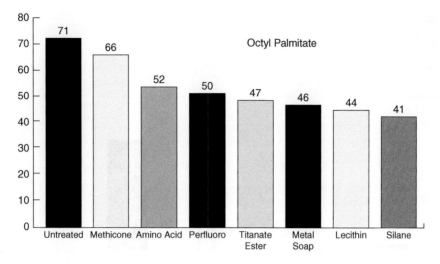

Figure 8.3   Oil absorption.

cosmetic have the same viscosity, stick hardness, or compressibility. The oil absorption of a pigment is determined by porosity, particle size, and affinity for the vehicle (wetting). The reader is referred to Table 6.1 to see the vast differences in oil absorption among the various types of pigments. To simplify this part of the formulating task, it may be desirable to reduce the level of absorption by those pigments that exhibit high oil absorptive powers and to surface treat pigments with compounds having functionality to optimize wetting in a given vehicle. Figure 8.3 shows the effects of various surface treatments on a single pigment: fine particle size mica with octyl palmitate as the oil. The test method employed was ASTM Method D-281-31, Oil Absorption of Pigments by Spatula Rub-Out, as discussed in Chapter 6. The untreated control has the highest oil absorption, while the silane treated sample has the lowest.

Suspension Viscosity: Figure 8.4 shows the effect of various types of surface treatments on the viscosity of a 30% suspension of the same mica and oil combination used for the oil absorption test. With particle composition and size constant, variations in suspension viscosity result from differences in wetting caused by surface group modification due to the various surface treatments. In this case, the perfluoro treated mica suspension actually has a higher viscosity than the untreated control, indicating that the treatment reduces wettability. Of those tested, titanate ester and silane treatments have the lowest viscosity suspensions, showing optimized wetting in octyl palmitate.

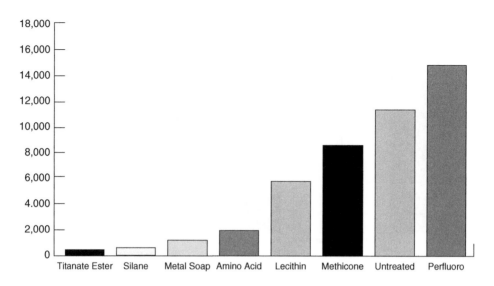

Figure 8.4   Suspension viscosity.

The remainder of the performance data are qualitative in nature, and will be presented as such, showing how the panelists rated the parameters in increasing order of preference.

Creasing: This is a qualitative ranking of the various treatment types based on their ability to prevent or minimize migration of eye shadow color into the crease between the eyelid and the eye socket. The results in Table 8.1 are listed in descending order by level of performance. The only two treatments to perform better than the untreated control are the perfluoro and the metal soap.

Oil Breakthrough: This property is, as its name suggests, a measure of how quickly sebum can be detected breaking through the film of a finished

Table 8.1   Wear results based on creasing.

Perfluoro
Metal soap
Untreated
Methicone
Amino acid
Lecithin
Silane
Titanate ester

Table 8.2 Wear results based on oil breakthrough.

Treatment type

Perfluoro
Untreated
Methicone
Metal soap
Amino acid
Lecithin
Silane
Titanate ester

Table 8.3 Wear results based on amount remaining.

Treatment type

Metal soap
Silane
Perfluoro
Untreated
Titanate ester
Methicone
Lecithin
Amino acid

cosmetic once applied on the skin. The evaluation was performed using a pressed powder eye shadow. The results in Table 8.2 are listed in descending order by level of performance. They show that only the perfluoro treatment is perceived by the panelists to be better in this regard.

Wear: This property is determined qualitatively by how much finished product remains on the skin after a set period of time. The results are shown in Table 8.3. Metal soap, silane, and perfluoro treatments can be seen to have increased retention of eye shadow. It should be noted that the silane treatment caused a negative effect: significant color change due to increased oil absorption.

Overall Appearance: This property is a measure of how the finished product appears after a period of wear, as compared to how it looked when first applied to the skin. The results, shown in Table 8.4, indicate that overall appearance after six hours' wear is improved over the control by metal soap, perfluoro, and Methicone treatments, despite less product remaining with the latter.

Table 8.4   Wear results based on overall appearance.

Treatment type

Metal soap
Perfluoro
Methicone
Untreated
Amino acid
Lecithin
Silane
Titanate ester

Intellectual Property Surrounding Surface Treatments

There have been many composition, process, and use patents issued on the subject of surface treatments, and while many of the earliest have expired, some remain in force today, as listed in Appendix B, based on a search completed by the author. It must be heavily stressed that a formulator who intends to use these materials must repeat this search (disclosure: the Delphion website aided the present one), as the patent landscape constantly changes as new patents are issued and others come to expiry.

Using Treated Pigments and Substrates

Though the use of treated pigments and substrates in cosmetic products is fairly straightforward, there are some factors that must be considered when formulating with them and incorporating them in the factory:

Surface Treatments vs. Coatings: Many of the treatments discussed in this chapter involve true surface bonding chemistry. They are applied to pigment and substrate surfaces by means of multi-step processes. It is possible to encounter coated materials in the market that are produced by a mechanical mixing process and, as a result, do not exhibit the same properties in a cosmetic process or a finished product. These coatings can be stripped off the surface of a pigment or substrate by friction during a milling process.

pH Use Range: Treated materials should be used in the pH range 5.0–8.0 because values outside this range can cause a disruption of the surface treatment.

Pulverization: Intense pulverization of materials treated with lauroyl lysine and other amino acids should be avoided because the integrity of the treatment can be compromised due to the heat sensitivity of the materials. Also, bulk powders containing materials treated with amino acids should be allowed to cool to ambient temperature after pulverization to prevent agglomeration.

Multiple Treatments: It is possible – and in some cases desirable – to use more than one surface treatment in the same formulation. Either multiple treatments of a single pigment or different but compatible treatments of each pigment in a blend will work. By using multiple treatments, the benefits of different treatment types can be realized in the finished product.

Partial Treatment Usage: It is not essential to treat every component of a formula. The number of components treated and the types of treatments used will depend on the type of cosmetic product involved and its intended function. For example, to prevent settling of inorganic pigments in nail polish, it is necessary only to treat those pigments, and none of the organic ones or other ingredients. Conversely, in order to produce good skin feel with a pressed powder, it is normal to treat all the dry components in the formulation.

Choosing Treatments

The whole subject of surface treatment is quite complex due to the myriad treatments available, coupled with the wide variety of pigments and substrates used in cosmetic products. The following are the main points to consider when going down the decision path:

Patents: This has been covered elsewhere in this chapter, but it is worth mentioning again. Due to the dozens of patents in force, the formulator must ensure that he/she is free to practice with a particular treatment in the finished cosmetic product to be formulated.

Application: Once a freedom to practice is confirmed, the next step is to determine what the application will be.

Property Improvement: The third step is to decide what property of the finished product is to be improved (e.g., skin feel, suspension, ease of application, compressibility, wear). The formulator can then research which treatments will be effective.

Antagonistic Effects: There is a possibility that a treatment type may have an adverse effect on another property of the intended product. The reader is referred to Figures 8.1–8.4 and Tables 8.1–8.4 for guiding information to help ensure that this doesn't happen.

Economics: As with all other aspects of formulating decorative cosmetic products, economics must play a role in the selection process. As more than one treatment type will usually serve the intended purpose, the formulator needs to review all of them to decide which will fit into the cost structure of the finished product. (See Chapter 5 for more discussion of economic factors in decision making.)

Reference

Faulkner, E.B. and Zavadoski, W.J. (1994). Pigment surface treatments: a bench chemist's guide. *Cosmetics and Toiletries* 109 (4): 69–72.

Chapter 9
Effect Pigments

All children love ice cream, and they love it even more when it has sprinkles on it. The adult equivalent to ice cream with sprinkles is a steak that comes to the table sizzling! In the world of color cosmetics, the sizzle on the steak or the sprinkles on the ice cream is the use of effect pigments to enhance the color properties of the absorption colors covered in the first eight chapters of this book. So, it is time to review the effect pigments, including natural pearl pigment, bismuth oxychloride, oxide coated micas and other composite pigments.

General Properties of Light

Before embarking on a discussion of effect pigments, it will first be necessary to review some general properties of light, as these are what make the effect pigments operate, producing their spectacular color enhancements. The reader is referred to Section 1.4 on "Color Physics" as a refresher on how light behaves. Terms relevant to the understanding of effect pigments are:

Incident Light: This is the light, composed of multiple wavelengths in the visible spectrum, that strikes a substance.

Coloring the Cosmetic World: Using Pigments in Decorative Cosmetic Formulations,
Second Edition. Edwin B. Faulkner. Edited by Jane C. Hollenberg.
© 2021 John Wiley & Sons Ltd. Published 2021 by John Wiley & Sons Ltd.

Absorption of Light: Based on the physical and/or chemical properties of a substance, the wavelengths of light striking it that pass into that substance are said to be absorbed.

Reflection of Light: Based on the physical and chemical properties of a substance, the wavelengths of light striking it that are not absorbed are said to be reflected from the surface of that substance.

Transmission of Light: This is the ability of a substance to allow light to pass completely through it, rather than being absorbed or reflected.

Angle of Incidence: This is the angle at which light strikes a substance.

Angle of Reflection: This is the angle at which light is reflected from the surface of a substance.

Specular Reflection: This type of reflection occurs when the angle of reflection is equal to the angle of incidence (Figure 9.1). The best example is the reflection created by a mirror: the sharpness of the image is directly related to the percentage of specular reflection the mirror is able to produce.

Diffuse Reflection: This type of reflection occurs when the angle of reflection is not equal to the angle of incidence, having reflection other than the "mirror image" just described. When all of the reflection is diffuse, it is not possible for the human eye to see any reflected image from the surface of the substance.

Refractive Index: This is a dimensionless number that relates the speed of light as it moves through a substance compared to that in a vacuum

Refraction of Light: This is a chemical property; when light passes from a medium of lower refractive index into a medium of higher refractive index,

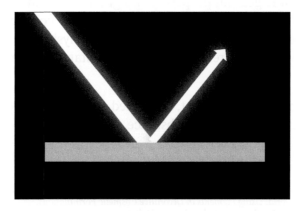

Figure 9.1   Specular reflection.

Figure 9.2   Glass prism.
Source: Courtesy of Color Techniques, Inc.

its speed is slowed and its path is bent away from the normal. The shorter
the wavelength of light, the greater the degree of bending of its path, which
separates white light into the component colors that make up the visible
spectrum. The best example of this phenomenon in nature is the refraction of sunlight by drops of rain to create a brilliantly colored rainbow. A
glass prism works on exactly the same "rainbow" principle (Figure 9.2).
Light passing through the prism is refracted, again showing the colors of
the visible spectrum.

Figure 9.3 illustrates all of the possible paths light can take after striking
the surface of a transparent or translucent substance.

When some or all of the properties just described come to into play, a
phenomenon called **light interference** can occur. That is, the substance
causes the light to be reflected, refracted, and transmitted, which **interferes**
with its journey through space, enhancing some wavelengths, while can-
celling others.

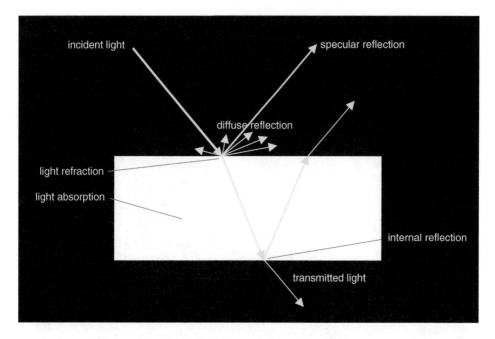

Figure 9.3   Interaction of light with a translucent substance.

General Light Interference Properties of Effect Pigments

With this basic understanding of light's behavior, it is time to move on to a discussion of the general properties of effect pigments. All of these types of pigments are constructed of various layers or platy particles, with different indices of refraction that allow for "interference" with incident light, causing maximum reflection, refraction, and transmission to create the play of color known as pearlescence. A very good example of these layers occurs in nature as mother of pearl, the interior of an oyster shell that produces the pearls so popular in modern jewelry. In mother of pearl, the materials of differing refractive indices that create the pearlescent effect are alternating layers of calcium carbonate and proteinacious material.

The key properties of effect pigments that allow them to work are as follows:

Platy Configuration: Flat particles having the high ratio of width to thickness known as aspect ratio provide a surface that maximizes reflection and minimizes light scattering from the particles' edges.

Table 9.1 Refractive indices of various materials.

Material	Refractive index
Air (vacuum)	1.00
Water	1.33
Mica	1.58
Guanine	1.80
Bismuth oxychloride	2.15
Titanium dioxide – anatase	2.55
Titanium dioxide – rutile	2.75
Iron oxide	2.96

Plate Thickness: There is an optimum thickness of the plate, dependent on the refractive index of the material, that maximizes light refraction and reflection to produce the optimum effect.

Smooth Surfaces: The smoother the surface of an effect pigment, the less diffuse reflection (light scattering) there will be. The amount of diffuse reflection created by an effect pigment is inversely related to the brilliance of its effect.

Transparency: The greater the degree of transparency exhibited by an effect pigment, the more refracted light is reflected through the platelet to maximize brilliance and create the illusion of depth.

Plate Length: There is an optimum length of the platy particles that maximizes specular reflection. The degree of specular reflection created by the plate is directly related to brilliance: the higher the reflection, the greater the brilliance.

Refractive Index: The higher the refractive index of an effect pigment, the more light splitting there will be, increasing the brilliance of the effect. Table 9.1 shows the refractive indices of various materials, most of which are used in the construction of effect pigments (see later).

Traditional Commercially Available Effect Pigments

With the fundamental properties of light and the interference properties of effect pigments in mind, a review of the traditional commercially available effect pigments follows.

Natural Pearl Pigment

Natural pearl pigment, also known as guanine, is made most commonly from the scales of the herring, caught in cold waters, but can also be derived from such fish as sturgeon and salmon.

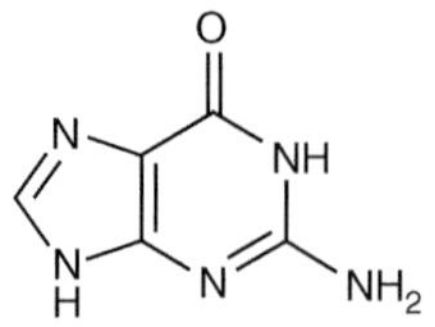

Figure 9.4
Guanine.
Source: Courtesy of
Sun Chemical Corp.

The first known synthesis and use of guanine was in seventeenth-century France, when a Catholic monk extracted it from fish scales to decorate rosary beads. In modern times, it became the first pearl pigment to be used in decorative cosmetics. The structure is shown in Figure 9.4 (Buxbaum and Pfaff 2005).

Guanine is produced commercially by extraction from fish scales with a strong solvent such as octane, then processing the resulting slurry of guanine/solvent through a series of phase transfers into vehicles that are compatible with the type of cosmetic products in which the pigment will be used. Typically, the vehicles are water, a water/surfactant blend for use in shampoos and lotions, or nitrocellulose for use in nail lacquer – the most common area of application in decorative cosmetics. Guanine can only be incorporated into fluid products as a dispersion due to the fragility of the particles. When dried, unless contained in a continuous film, the crystals will turn to dust with the slightest pressure.

Guanine offers a unique soft luster, but became seldom used in cosmetics due to high cost, manufacture using environmentally objectional solvents, and concern among some consumers over the animal source. There is, however, some renewed interest in its use among formulators following the currently popular trend toward the use of natural raw materials in personal care products.

Bismuth Oxychloride

Bismuth oxychloride, the first commercially available synthetic pearl pigment suitable for cosmetic use, was introduced to the cosmetic industry in the late 1950s. Pigmentary bismuth oxychloride is produced by dissolving mined bismuth ore in concentrated hydrochloric acid, forming bismuth trichloride ($BiCl_3$), which is then hydrolyzed to its final chemical form, $BiOCl$.

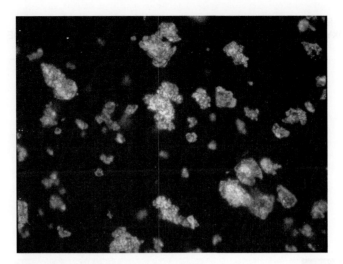

Figure 9.5   Bismuth Oxychloride.
Source: Courtesy of Sun Chemical Corp.

Unlike guanine, bismuth oxychloride can be processed as a dry pigment or, by means of phase transfer, put into suitable vehicles for use in decorative cosmetics, the most common being nitrocellulose for nail polish.

While guanine is essentially only available in a single particle size range and appearance, BiOCl is available in a variety of particle sizes, which produce different effects. Figure 9.5 is a photomicrograph of BiOCl crystals that shows their typical bipyramidal structure: platy but thicker in the center. Generally speaking, for all effect pigments, reflection increases and coverage decreases as the particle size of the crystal or plate increases. Some typical particle size ranges and their effects for BiOCl are:

Fine Grade (1–20 μm): Pigments in this range are mostly offered in dry powder form, for use in powder and anhydrous color cosmetics, exhibiting excellent coverage and a smooth luster.

Medium Grade (5–50 μm): Pigments in this range have medium coverage and greater effect. They are available in powder or dispersion form, the degree of luster being dependent on the crystal shape. Dispersions of regular, intact BiOCl crystals in this range are a replacement for the soft luster effect of natural pearlescence.

Large Grade (20–80 μm): Pigments in this range exhibit high luster and lowest coverage. Nitrocellulose lacquer dispersions of these larger, flat crystals create a dramatic mirrorlike effect in nail lacquer.

BISMUTH
OXYCHLORIDE
1–20 µM

BISMUTH
OXYCHLORIDE
5–50 µM

BISMUTH
OXYCHLORIDE
10–80 µM

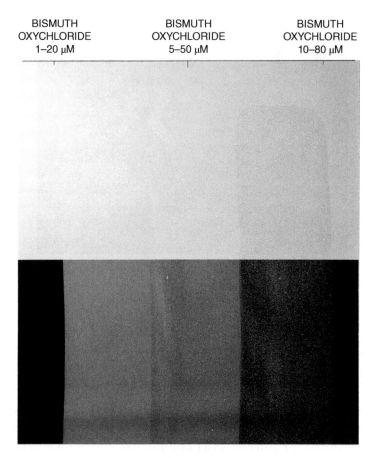

Figure 9.6   BiOCl: display of three particle sizes.
Source: Courtesy of Sun Chemical Corp.

Figure 9.6 is a comparative display of these three particle size ranges as drawdowns over black and white substrates. Color shows on the white surface, while coverage and luster are evaluated over the black portion of the card.

The advantages of using BiOCl are as follows:

Coverage/Luster Balance: Like guanine, BiOCl provides good luster while also offering good coverage on the skin. It is the closest of the synthetic effect pigments in these respects to guanine, at considerably less cost.

Tactile Properties: BiOCl has a very good feel when rubbed on the skin, so any of the grades will enhance the feel of a finished decorative cosmetic product. In fact, in many instances, small particle size BiOCl pigments are used specifically to improve feel, not for their effect.

Smooth Luster: The type of luster exhibited by BiOCl is very smooth as compared to the sparkle luster of many of the metal oxide coated effect pigments (see later). In formulas where this smooth luster is desired, BiOCl becomes the effect pigment of choice.

Coverage: Again, with reference to the metal oxide coated pigments, BiOCl exhibits much better coverage but is also more expensive. The formulator must factor the better coverage into the cost equation, not simply compare the price per kilo of the BiOCl versus the alternatives.

Compressibility: BiOCl is a very compressible material, so it can facilitate binding in compressed powder formulas.

There are also some disadvantages with the use of BiOCl, which are as follows:

Light Stability: BiOCl is inherently not stable to light. If used in a product displayed in clear packaging, it will darken even when exposed to fluorescent or incandescent light in a retail environment. There are light stabilized grades available that will perform better when exposed to visible and ultraviolet light. However, even with these, it is critical that the formulator runs the proper light stability tests to ensure the finished product will perform satisfactorily in the packaging in which it will be marketed.

Fragility: BiOCl crystals, whether dry or dispersed in paste forms, are fragile. When processing the higher luster grades, any type of harsh mixing or milling must be avoided. These are truly stir-in pigments and should always be added to formulas late in the process to avoid any destruction of the crystal. To enable adequate dispersion without loss of effect, the higher luster, larger particle size grades are sold by their manufacturers as pre-made pastes, particularly nitrocellulose lacquer for use in nail enamel. The powder forms of BiOCl used for functional purposes rather than visual effect are less sensitive to high shear mixing. As with any product, the formulator should still evaluate variations in manufacturing procedures for possible changes to appearance or texture. A self-prepared, gently stirred premix in a portion of the formula oil when using powder grades in lip products or cream eye shadows may be advisable to aid dispersion without excessive force.

Density: The specific gravity of BiOCl is 7.73. While high density may be an advantage in pressed powder formulations, suspension in liquid vehicles is difficult.

Color Palette: The BiOCl crystal is only available as a white (silver) pigment. There are pigments available that are combinations of BiOCl with standard absorption pigments, in order to exhibit other color shades. The most frequently used are intimate blends with Iron Blue, Carmine, and Chromium Greens.

Economics: Even when the coverage ratio comparison with other silver-white effect pigments is taken into consideration, BiOCl pigments are more expensive. As a result, they are commonly only used where their advantages can be realized in the decorative product into which they are formulated. For reference, in the United States the dry form of BiOCl sells in the range of $85.00/kg, while a diluted paste counterpart sells for about $55.00/kg at a pigment content of 25%. Prices in other parts of the world will vary based on transportation costs, duties, and currency exchange rates.

BiOCl is used in a wide variety of decorative cosmetics, but the major uses are in powder and cream eye shadows, lip products, and nail lacquer. In eye products, the dry types are used, enabling the formulator to achieve a smooth pearlescent look while enhancing the feel and adhesion of the product when applied to the user's skin. In addition, BiOCl improves compression in pressed powder formulations.

In nail polish, BiOCl is most commonly employed in the paste form, in which the larger, more fragile BiOCl crystals provide a better luster than does the dry form of the pigment. When the product is applied to the nail, the desired smooth swirl effect can easily be seen, as the flat particles orient parallel to the surface of the nail to maximize reflection.

Although BiOCl was originally useful as a pearlescent pigment, other materials were investigated in order to develop a more light-stable, less fragile alternative. Titanium dioxide has the required high refractive index, but single crystals with a suitable platy configuration were never successfully manufactured. In 1963, DuPont patented the coating of transparent mica platelets with thin films of titanium dioxide to create particles with high refractive index, high aspect ratio, and transparency – the properties required for pearlescence (Buxbaum and Pfaff 2005). This technology was then licensed to the Mearl Corporation and E. Merck, which were the first companies to offer pearlescent titanium dioxide coated mica pigments to the cosmetic industry in the 1960s. Both companies continued development of the technology to achieve a wide range of effects and colors.

Editor's Note: The majority of effect pigments currently used in color cosmetics consist of flake pigments composed of coatings of titanium dioxide and other cosmetic color additives on transparent cosmetically acceptable bases. The next section describes the composition and effects that can be achieved with the original mica substrate in detail. The principles and processes used to create mica based effect pigments are the same as those involved in the production of effect pigments on the additional substrates used for cosmetic effect pigments, as discussed later in this chapter.

Oxide Coated Micas

By the early 1970s, metal oxide coated mica pigments became the most important class of effect pigments used in decorative cosmetics These are composite mixtures using refined natural muscovite mica as a base onto which an absorption pigment, commonly titanium dioxide and/or iron oxide, is deposited.

The mica refining process has two purposes. First, as the mica is mined, after removal from the earth, contaminating debris must be expelled. Second, the mica is classified into various particle size ranges; the impact of this will become clear later in the chapter. Figure 9.7 shows mica as it comes from the mine (lower image) and after it has been refined (upper image).

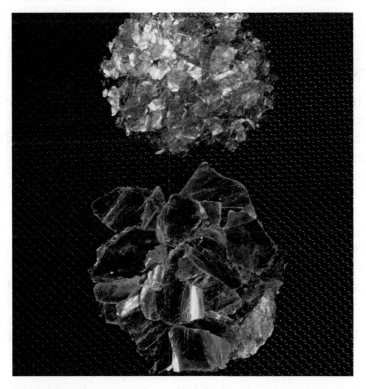

Figure 9.7   Top: Refined mica. Bottom: Raw mica.
Source: Courtesy of Sun Chemical Corp.

Figure 9.8   Titanium dioxide on mica carrier.

After refinement, the mica is completely coated with a metal oxide layer as shown in the schematic diagram in Figure 9.8. The process consists of precipitation of titanium hydroxide from a solution of titanyl sulfate or titanium tetrachloride onto the surface of refined mica, followed by filtration, and finally high temperature calcination to form a film of titanium dioxide on the flat mica surface.

As with BiOCl, the mica based pearl pigments are available in a variety of particle size ranges, which are generated during the mica refining process. These pigments follow the general rule that as the particle size increases, so does the luster or sparkle, but this is at the expense of the coverage. The commonly available particle size ranges for effect pigments based on mica and other substrates can be described as follows:

Extra Fine Grade (<15 µm): Pigments in this range exhibit a satin luster and have a very high degree of coverage.
Fine Grade (5–20 µm): Pigments in this range produce a low level of luster and maintain a high degree of coverage.
Medium Grade (10–50 µm): Pigments in this range exhibit a medium level of luster and have a good degree of coverage. This grade is the most commonly used, due to the balance of luster and coverage, which provides good value in use.
Large (Sparkle) Grade (10–100 µm): Pigments in this range exhibit a sparkle effect with reduced coverage.
Extra Large (Sparkle) Grade (20 to >150 µm): Pigments in this range show a more dramatic sparkle effect with little coverage.

Note: In the United States, mica for drug and cosmetic use is regulated as a color additive, subject to the specifications contained in the Code of Federal Regulations (CFR), as discussed in Chapter 2.

There is a particle size limit of 150 µm for mica, to which all micas or mica based effect pigments sold for drug or cosmetic use in the United States must conform. Effect pigments based on substrates not classified as color additives do not have a legal particle size limit, thus larger flakes can be used for a glittering appearance. Considering the potential for skin or eye irritation caused by large particles, the formulator should follow the pigment manufacturer's usage recommendations and evaluate the safety of any particulate containing product in the end use.

Figure 9.9 shows the effect of particle size on appearance with displays of the extra fine through sparkle pigment types over both black and white

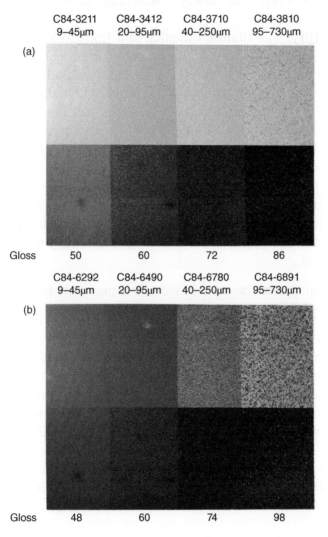

Figure 9.9 Oxide coated mica four-pigment display: (a) coated with titanium dioxide and iron oxides; (b) coated with iron oxides.
Source: Courtesy of Sun Chemical Corp.

backgrounds. The pigments in Figure 9.9a consist of mica coated with tita-nium dioxide to create pearlescence and with iron oxides to achieve the gold masstone, while those in Figure 9.9b are mica coated with only iron oxides, which produce both the luster and the masstone. (For more information on colored effect pigments, see later.)

The advantages of using oxide coated micas are as follows:

Color Palette: Unlike guanine and BiOCl, oxide coated effect pigments are available in a very wide variety of whites and colors. Their composition will be covered later.

Light Stability: The titanium dioxide and iron oxide coated effect pig-ments have excellent light stability in most cosmetic applications. Resis-tance to color change is so good, in fact, that with additional protective coatings, they are used in automotive coatings, where this property is criti-cal, as the object itself (car, truck, etc.) is under constant exposure to light, both natural and artificial. The light stability of other colorants added to the coating is characteristic of the particular material used.

Economics: These pigments offer by far the best value for money of the three effect pigment types discussed so far. Even though they have less cov-erage at equal particle size compared with BiOCl, the mica based effect pigments still provide the best economic choice when used in decorative cosmetics. Additionally, effect, not coverage, is usually the primary prop-erty of interest.

The advantages of the mica based effect pigments far outweighs their disadvantages, which are as follows:

Fragility: Although not as fragile as guanine and BiOCl, mica is easily frac-tured and must not be subjected to any type of harsh milling or grinding. Rather, the mica and other substrate based effect pigments are true "stir-in" colorants that can be dispersed in most products with simple mixing.

Transparency: As previously mentioned, metal oxide coated effect pig-ments do not exhibit the same degree of coverage as BiOCl at equal particle size.

Graininess: Even the extra fine platy pigment based effect pigments have a perceivably more particulate luster compared to both BiOCl and guanine. A grittier skin feel is also noticeable, especially with the larger particle size grades.

Color Effect of Oxide Coated Micas

As noted earlier, one of the key advantages of effect pigments made by coating a variety of transparent bases with metal oxides is the ability to achieve a wide range of colors and effects. The two factors that determine a pigment's color are the type of oxide used to coat the substrate and the thickness of that oxide layer. For the deposited high refractive index oxide layer, titanium dioxide (TiO_2) and/or iron oxide are the materials of choice.

As mentioned at the start of the chapter, multiple layers of materials with differing indices of refraction allow for "interference" between light rays reflected, refracted, and transmitted at their many interfaces. In the case of titanium dioxide, those pearls with the thinnest coating layers are perceived by our eyes as having a white masstone/absorption color, observable at all viewing angles, AND a white reflection color, observable at the spectral angle. As the oxide layer's thickness increases, precise control of the interaction between light rays can create reflection colors due to light interference. All colors of the rainbow are possible. The branch of physics called optics describes the derivation of these thin film interference effects. Briefly:

Electromagnetic radiation, including visible light, travels in waves (see Figure 9.10). Red light has the longest wavelength, violet the shortest.

The light refracted by passing through the titanium dioxide layer is divided into its component colors (see the prism in Figure 9.2).

When the refracted light is reflected from the boundary between the titanium dioxide layer and the mica platelet, the refracted rays interact with the light reflected from the top of the titanium dioxide layer (see Figure 9.11).

Wavelengths that emerge in phase are enhanced, intensifying their color. This is called **constructive interference** (see Figure 9.12).

Wavelengths that are out of phase are effectively cancelled, weakening their color. This is called **destructive interference** (see Figure 9.13).

The single color reflection, called **interference color** or **reflection color**, is determined by the thickness of the oxide layer. **Optical thickness**, which is the layer thickness multiplied by the refractive index, quantifies

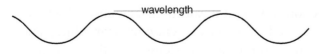

Figure 9.10   Light waves.

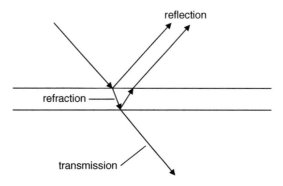

Figure 9.11   Simultaneous light reflection and refraction.

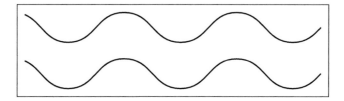

Figure 9.12   Light waves in phase.

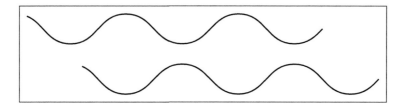

Figure 9.13   Light waves out of phase.

the effective influence of the high refractive index material on the light passing through the layer.

Table 9.2 lists the optical thicknesses of the titanium dioxide layer that create different colors.

The interference color, observed at the spectral angle, is caused entirely by light reflection, not light absorption. The absorption color of such titanium dioxide coated substrates is white or near-white, since both materials normally reflect almost all wavelengths of light. No colored pigment is present. The transmission color is the complement of the reflection color.

Iron oxide is another cosmetic colorant with a high refractive index that can be coated on platy substrates to produce interference pigments. The

Table 9.2   Interference colors.

Titanium dioxide layer thickness (nm)	Color
100–140	White
140–200	Gold
200–230	Orange
230–280	Red
280–310	Violet
310–360	Blue
360–390	Green

Table 9.3   "Metallic" colors.

Iron oxide layer thickness (nm)	Color
100–140	Bronze
140–170	Copper
170–200	Red
200–230	Maroon

color of effect pigments created with iron oxide is the result of the combination of the interference (reflection) color, due to the thickness of the oxide layer, with the reddish brown absorption color of the iron oxide itself. The different thicknesses of the iron oxide layer leading to different colors are shown in Table 9.3. Several bright colors, commonly called metallics, are available: a bronze (gold interference), a copper (orange red interference), a red (red interference), and a maroon (blue interference).

A gold metallic color can be produced by coating the platy substrate with both titanium dioxide and iron oxide (see Figure 9.9).

The color palette can be further expanded by depositing other absorption pigments onto interference colors, thereby creating a three-layer coating (see Figure 9.14). These three-layer pigments can be called "enhanced interference" colors, exhibiting interference (reflection) color at the spectral angle and absorption color at other viewing angles. The common absorption colors used in these pigment mixtures are ferric ferrocyanide, carmine, and chromium oxide green. Depending on which is used with what type of interference color, different effects can be obtained. If the absorption color is the same as the interference color, an intensified interference color results; an example is the deposition of carmine on to a red interference color to make

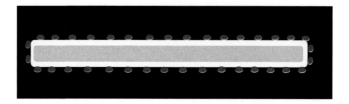

Figure 9.14   Colored TiO$_2$/mica pigment.

a brilliant red pigment. If the absorption color is different from the interference color, a dramatic dichroic (two tone) angle dependent effect results; an example is the deposition of ferric ammonium ferrocyanide onto a red interference color to produce a pigment that appears red at the spectral viewing angle and blue at other angles.

Usage of Oxide Coated Pigments in Decorative Cosmetics

Unlike the absorption colors, which are restricted in their use by stability and/or regulatory issues, the oxide coated effect pigments find, to one extent or another, usage in virtually the entire gamut of decorative cosmetic products. They are used extensively in lipstick, lip gloss, eye shadow, eyeliner, mascara, and nail polish applications, and even find some use in liquid makeup. The regulatory status and stability characteristics of the enhanced interference and two-tone pigments are, however, affected by the absorption color used. (Some regulatory cautions will be addressed later in the chapter.)

While these pigments do not suffer from as many stability issues as the absorption pigments, there are some cautions that must be noted:

Combination with Absorption Pigments: By their nature, effect pigments are transparent and must therefore be used with transparent absorption colors or with low levels of opaque colors. If this guideline is not followed, the large quantity of effect pigment needed to produce the desired pearlescent effect can negatively influence the physical and economic characteristics of the finished decorative product.

Fragility: Although mentioned previously in this chapter, it bears repeating: all of the effect pigments are very fragile and must not be subjected to any dispersion or grinding processes. They are all stir-in products, which is one of their advantages, because they do not have to undergo separate dispersion steps as do the absorption pigments.

Alkaline Stability: Enhanced interference or two-tone effect pigments made with ferric ferrocyanide are not stable in alkaline systems, because the iron blue is not stable in those systems.

Acid Stability: Enhanced interference or two-tone effect pigments made with carmine are not stable in acid systems, because the carmine is not stable in those systems.

Heat Stability: Enhanced interference or two-tone effect pigments made with carmine are not stable above 60 °C, because the carmine is not stable above that temperature threshold.

Settling: The substrates used for the effect pigments have relatively high specific gravity compared to cosmetic vehicles, so they will tend to settle in liquid systems unless a gellant is added.

Regulatory Considerations When Using Oxide Coated Pigments in Decorative Cosmetics

I'll keep this brief, and refer the reader back to Chapter 2 for a more in-depth look at regulatory considerations.

With regard to the **organic lakes**, caution must be used when formulating effect pigments that have a layer of organic lake deposited on to them. Under US regulations, mica and the other substrates used for effect pigments are not listed as approved substrates for the formation of lakes of the certified organic colors. Instead, finished certified lake pigments may be physically bonded to finished effect pigments, resulting in a legal mixture of the two. The formulator must also ensure that the organic lake is permitted for the intended use of the product in the market where it will be sold. For example, an effect pigment with a layer of D&C Red 7 Ca Lake deposited on to it would not be permitted for use in eye makeup sold in the United States.

As far as **exempt colorants** are concerned, caution must likewise be used when formulating effect pigments that have a layer of exempt color additive deposited on to them. The formulator must ensure that the exempt color additive is permitted for the intended use of the product in the market where it will be sold. For example, an effect pigment with a layer of ferric ferrocyanide or chromium oxide green deposited on to it would not be permitted for use in lipsticks sold in the United States.

Finally, the United States Food and Drug Administration (FDA) specification for natural mica used in drugs and cosmetics has a particle size limit of 150 µm, so effect pigments based on mica with particle sizes above this level cannot be used. This regulation does not pertain to effect pigments based on borosilicate, alumina, silica, or fluorophlogopite.

Other Effect Pigments

Over the last 20 years or so, several different substrates for effect pigments have been commercialized using the same layered mixture technology employed in the manufacture of the oxide coated micas. These substrates can be the base for white, interference, enhanced interference, two-tone, and metallic pigment types. The most significant are as follows:

Synthetic Mica Effect Pigments: These pigments are similar to oxide coated natural mica pigments, but their base is synthetic fluorophlogopite, which is a synthetic mica. Because the mica is synthetic, its platelets are smoother and do not contain the high mineral content of natural mica, giving them a much brighter, clean white appearance. In pressed powder formulations, the fluorophlogopite based effect pigments compact more easily than do the mica pigments, forming stronger cakes. In addition, because these pigments are not restricted to $<150\,\mu m$ particle size, brilliant sparkle effects can be obtained with pigments that range up to $750\,\mu m$.

Borosilicate Effect Pigments: The transparency of borosilicate glass used as a base in effect pigment mixtures results in unparalleled depth and brilliance of color effects. As with synthetic mica, particle sizes up to $750\,\mu m$ are possible, which produces spectacular sparkle effects. I repeat: in the absence of regulatory restrictions, it is up to the cosmetic formulator to test the safety of products containing large particle size effect pigments, particularly around the area of the eye.

Silica Based Effect Pigments: Pigment mixtures utilizing silica as a base exhibit much cleaner masstones than conventional oxide coated mica mixtures and introduce the possibility of color travel in decorative cosmetic products. Successive layers of titanium dioxide and silica provide multiple transitions between materials of differing refractive indices, and so multiple opportunities for refraction and reflection as light passes through the coating. "Color travel" refers to the ability of these pigments to exhibit multiple colors as they are viewed from multiple angles, even in the absence of any absorption pigment.

Alumina Based Effect Pigments: Pigment mixtures utilizing alumina as a base can produce color travel and/or help mask imperfections in the skin.

Iron Oxide Coated Pigments: While seemingly similar to "metallic" oxide coated micas, these patented pigments produce bright clean colors that are not possible with the conventional iron oxide coated substrates.

Organic Lake/Oxide Coated Micas: Natural and synthetic mica coated pigments are also coated with organic lakes. The pigments resulting from this process exhibit very colorful bright pearlescent colors that are quite easy to disperse. It should be noted, as already mentioned, that lakes are not formed on the mica substrate, because mica and fluorophlogopite do not appear on the FDA's list of approved substrates for lake formation in 21CFR82.51 and 82.1051. Instead, pre-made organic lake pigments are adhered to the effect pigment surface.

Economics of Oxide Coated Micas and Other Effect Pigments

Before moving on to the discussion of specialty pigments in Chapter 10, it is important to note that there is a very wide range of prices for the effect pigments described in this chapter. The natural mica based white pigments are reasonably affordable, while those based on alumina and borosilicate glass are quite expensive. As a result, the formulator must be aware of the per kilo cost of the pigment he/she intends to use and the percentage that will be required based on its particle size before starting work on the bench. As a guide, Table 9.4 lists the market prices for the types of pigments reviewed in this chapter. Due to the different sources, freight costs, and duty rates throughout the world, as well as fluctuations in currency values, the prices in this table are expressed as multiples of the white natural mica based pearls, which are shown as 1.00.

Table 9.4   Relative prices of effect pigments.

Effect pigment type	White	Interference, colored	"Metallic"	Bright colors
Natural mica substrate	1.0	2.0–3.0	1.8–2.4	N/A
Fluorophlogopite (synthetic mica) substrate	3–5	5–7	5–7	N/A
Borosilicate (glass) flake substrate	9–12	10–14	10–14	N/A
Silica flake substrate	18–22	18–22	N/A	N/A
Alumina flake substrate	18–22	18–22	N/A	N/A
Iron oxide coated	N/A	N/A	N/A	7–9
Organic lake coated	N/A	N/A	N/A	10–12

The information in Table 9.4 is an approximation of the relative costs of the different types of effect pigments. The prices are based on average particle size, which in most cases is in the 10–50 µm range. It follows that finer pigments will be priced lower and larger ones will be priced higher.

Reference

Buxbaum, G. and Pfaff, G. (eds.) (2005). *Industrial Inorganic Pigments*, 3e. Wiley-VCH Verlag GmbH & Co.

Chapter 10

Specialty Pigments

This chapter will focus on specialty pigments. The pigments in this category tend not to be used in large volumes, because of regulatory restrictions, handling and compatibility issues, and cost, or due to the fact that their effect is only needed in certain instances. Despite this limited usage, these pigments bring spectacular, unique effects to the products in which they are incorporated, and are therefore an extremely important class of color additives for the decorative cosmetic industry.

Metallic Pigments

Chapter 9 covered a class of effect pigments referred to as metallics, which are bronze and copper colored pigments composed of iron oxide or titanium dioxide coated natural and synthetic substrates. The metallic pigments discussed here are truly metallic – essentially the pure metals aluminum, copper, and bronze, as represented in Figures 10.1 (bronze and copper) and 10.2 (aluminums). Their mechanism of color display is different from that of both the absorption and the effect pigments.

The absorption colors operate by selectively absorbing and diffusely reflecting different wavelengths of light, as shown in Figure 10.3. The circles represent particles of pigment on the surface of the skin. The long arrows show incident light striking the surface of the pigment and being

Coloring the Cosmetic World: Using Pigments in Decorative Cosmetic Formulations,
Second Edition. Edwin B. Faulkner. Edited by Jane C. Hollenberg.
© 2021 John Wiley & Sons Ltd. Published 2021 by John Wiley & Sons Ltd.

Figure 10.1   (a) Bronze pigment. (b) Copper pigment. (c) Copper pigment.
Source: Courtesy of Eckart.

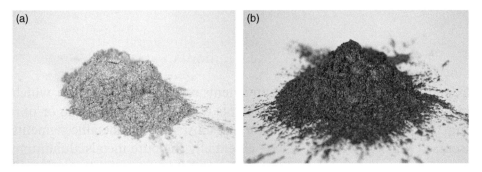

Figure 10.2   (a) Aluminum pigment. (b) Aluminum pigment (blue).
Source: Courtesy of Eckart.

reflected at diffuse angles. The short arrows penetrating into the pigment demonstrate light being absorbed.

Effect pigments based on metal oxide coated substrates operate by selectively absorbing, reflecting, refracting, and transmitting light, as described in detail in Chapter 9 and shown in Figure 10.4. The arrows show incident light being reflected at both the specular and diffuse angles and being transmitted through the crystal.

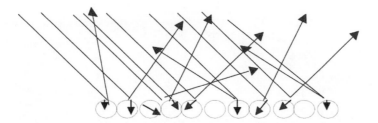

Figure 10.3   Absorption pigment light reflection.
Source: Courtesy of Eckart.

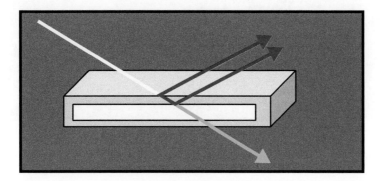

Figure 10.4   Effect pigment light interactions.
Source: Courtesy of Eckart.

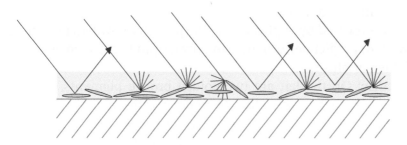

Figure 10.5   Metallic pigment light reflection.
Source: Courtesy of Eckart.

The metallic pigments, on the other hand, produce their color effects by reflecting the entire wavelength range of incident light, a very high degree of which is at the specular angle. This produces a bright metallic look like that seen in jewelry. Figure 10.5 shows this mechanism, where the flat, plate-like particles are the metallic pigment and the arrows are incoming reflected light, most of which is at the specular angle.

Figures 10.6 and 10.7 are comparisons among two types of metallic pigments and two types of oxide coated effect pigments. Masstones and

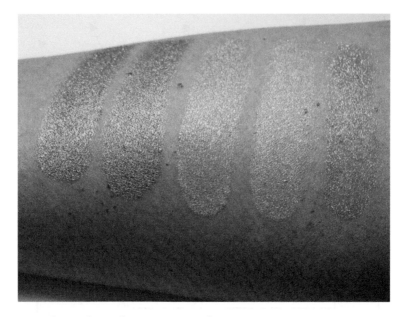

Figure 10.6   Effect and metallic pigment powder comparisons.
Source: Courtesy of Eckart.

reflection colors of all are gold, but composition influences luster, color, and coverage. Readers may find it instructive to see the pigments as they appear in simulated use.

Effects observed by application of the pure dry powders to an inner arm (Figure 10.6) predict performance in powder products, such as eyeshadow, highlighter, or blush.

Metallic: On the left are two golden bronze pigments, composed of bronze powder coated with silica. Their effect is the smooth sheen and high coverage created by the metal flakes. The particle size of the far left sample is the larger of the two, exhibiting stronger reflection and less coverage.

Synthetic Mica Base: The middle pigments consist of oxide coated synthetic mica (fluorophlogopite). Both are composed of fluorophlogopite platelets coated with titanium dioxide, iron oxides, and tin oxide to give a gold masstone combined with a gold reflection. The left middle pigment is an example of a newer technology, showing a more intense color and higher coverage compared to the traditional coated synthetic mica to the right.

Borosilicate Base: The pigment on the far right is composed of calcium sodium borosilicate platelets coated with titanium dioxide, iron oxides, silica, and tin oxide. The borosilicate base allows the highest transparency and light transmission to create a maximum effect.

Figure 10.7   Effect and metallic pigment drawdown comparisons.
Source: Courtesy of Eckart.

Drawdown of nitrocellulose dispersions (Figure 10.7) specifically pre-
dicts performance in nail lacquer, but also shows effect differences to be
expected in other formulations in which the powder is wet into a liquid vehicle, such as the oils used in lip color. Pigments shown are similar to those in
Figure 10.6, only in reverse order.

Borosilicate Base: Sparkle effect and transparency are maximized in
the far left of the drawdown.

Synthetic Mica Base: The pigment in the left middle dispersion is an
example of traditional oxide coating technology on fluorophlogopite. The
more intense color of the newer coating technology to the right shows to
good advantage in dispersion form.

Metallic: On the right, golden bronze dispersion has the coverage and
polished metal appearance valued in high fashion nail lacquers.

The appearance of metallic pigments, like that of effect pigments, is
controlled by several factors, including particle size, shape, distribution,

and orientation. Of these factors, the influence of particle size dominates, impacting the metallic character and optical properties of the pigment. With respect to the former, coarser particles have a higher surface to edge ratio, which produces fewer scattering centers, resulting in a higher degree of specular reflection. This gives these pigments a more brilliant metallic character. Finer particle pigments have a lower surface to edge ratio, leading to more scattering centers and thus a higher degree of diffuse reflection, which yields more of a soft, less brilliant metallic character. The brightness of the metallic pigments increases as their particle size increases, while the coverage does the reverse, decreasing as the particle size increases.

The general particle size ranges are as follows:

Fine Grade (5–50 μm): Brilliant effect with excellent coverage.
Medium Grade (15–70 μm): Bright effect with good coverage.
Large Grade (20–95 μm): Sparkle effect with fair coverage.

Aluminum Pigments

Traditionally, aluminum pigments were manufactured by processing aluminum particles through a ball mill in the presence of a solvent, followed by classification through a sieve, then solvent removal by means of a filter press, before final processing into dry or dispersion forms. This production technique produced particles that are best described as having a cornflake structure, with edges measuring in the 100–500 nm range (Figure 10.8).

More recently, a novel process for the manufacture of aluminum pigments was developed, in which aluminum is vaporized and deposited on

Figure 10.8 Cornflake pigment, electron photomicrograph.
Source: Courtesy of Eckart.

to a polyester film. The pigment is then stripped from the film, classified, and processed into dry or dispersion forms. The particle produced by this method has edges that measure in the 30–50 nm range.

The difference in the edges between the two products makes a dramatic difference in the effect they produce. The conventional cornflake product's edges cause more light scattering than those of the vapor deposition product, which results in more diffuse reflection, and therefore a more satin metallic effect. The vapor deposition product produces an almost mirror-like reflection effect. In the electron micrographs in Figures 10.9 and 10.10, the difference between the two types is readily evident; it can be seen how much thinner the vapor deposition pigments are.

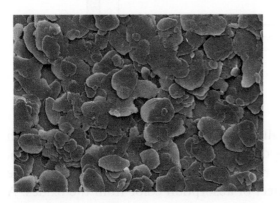
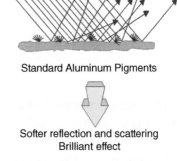

Standard Aluminum Pigments

Softer reflection and scattering
Brilliant effect

Figure 10.9 Cornflake pigment, photomicrograph.
Source: Courtesy of Eckart.

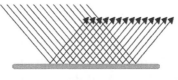

Uniform reflection
Very brilliant mirror-like effect

Figure 10.10 Vapor deposition pigment, photomicrograph.
Source: Courtesy of Eckart.

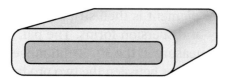

Figure 10.11   Aluminum hydroxide
coated aluminum.
Source: Courtesy of Eckart.

Figure 10.12   Ferric ferrocyanide/silica coated aluminum.
Source: Courtesy of Eckart.

The latest generation of metallic aluminum pigments is a series in which a layer of one or more oxides are deposited on the surface of the aluminum. The combination of the two materials results in pigments with the brilliance of aluminum and the added effect of absorption color from the oxides. A layer of aluminum hydroxide alone (Figure 10.11) imparts a champagne color to the aluminum, while a combination of silica and ferric ferrocyanide (Figure 10.12) brings a brilliant blue to the effect party.

Aluminum Pigment Regulations

Table 10.1 shows the regulatory status of aluminum usage in the major world markets. It is regulated in the United States, the European Union, and China. As it does not fall into the category of Coal Tar Colorants, it is not regulated in Japan.

In the United States, the major regulatory drawback to aluminum is the fact that it is not permitted for use in ingested cosmetics (i.e., lipstick). In addition, there is a particle size limit of 74 µm. In the European Union, aluminum is a food colorant, so it must meet the food color specifications found in E173, which is the food color listing for aluminum in the European Union. There is no such requirement in China, so it only has to meet the criteria set out in the Safety and Technical Standards for Cosmetics (2015 version).

Table 10.1   Aluminum pigment regulations.

Area of use	United States	European Union	China	Japan
Mucous membrane – oral	Not permitted	Permitted	Permitted	Not regulated
Eye area	Permitted	Permitted	Permitted	Not regulated
External	Permitted	Permitted	Permitted	Not regulated
Rinse off	Not applicable	Permitted	Permitted	Not regulated

The formulator must exercise caution when using two-layer pigments to insure that the absorption color used meets the regulatory standards of the market where the decorative cosmetic will be sold. For example, an aluminum pigment coated with D&C Red No. 7 Ca Lake could not be used in an eye-area product designed for sale in the United States, because D&C Red No. 7 Ca Lake is not permitted for eye-area use in the United States.

Using Aluminum Pigments

Aluminum pigments are used in a wide variety of anhydrous decorative cosmetics, including eye shadow, lipstick (except in the United States), mascara, and nail polish. The conventional cornflake structured pigments find use in all of these products, while the vapor deposition pigments are used largely in nail polish, where their mirrorlike finish has the most value. When using them in this application, the following guidelines should be observed in order to maximize the optical brilliance effect. They should be used:

- in low viscosity systems, typical of nail lacquer;
- with low resin content;
- at a low pigment load; and
- at a low film thickness, typically 0.1–0.3 mils.

The aluminum pigments do have some stability issues that must be considered when formulating them into decorative cosmetic products. First and foremost, they will react with water to form aluminum hydroxide, liberating hydrogen (H_2) in the process. This can cause obvious problems, so they should not be used in aqueous systems as a neat pigment. There are encapsulated grades on the market, where the aluminum flakes are completely surrounded by a layer of silicone dioxide (SiO_2). These materials can be used in aqueous systems, but stability should be studied carefully.

Caution must also be observed with two-layer pigments to insure that the absorption pigment deposited on to the aluminum is stable in the system

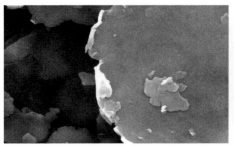 

Figure 10.13   Aluminum pigments deformed by high shear.
Source: Courtesy of Eckart.

where the composite pigment will be used. For example, an aluminum pigment coated with ferric ammonium ferrocyanide cannot be used in alkaline
systems, as the blue is not stable above pH 7.

The major use issue with aluminum pigments is their fragility; they are
very susceptible to deformation when exposed to any high shear forces. In
this respect, they are similar to the effect pigments, but in this case, the flakes
will tear and bend, resulting in an increase in light scattering, diminishing
their metallic effect. Also, in the case of the encapsulated pigments, free
metal surfaces can be exposed, increasing the risk of reaction with water.
Due to this deficiency, aluminum pigments should be treated with as much
or more care than effect pigments when being incorporated into any decorative cosmetic product. They must not be exposed to any harsh milling or
grinding, and, where possible, they should be put into the product at the end
of the process. Figure 10.13 shows two SEM photos, magnified 16 000×, of
aluminum pigments that have been exposed to high shear forces.

Copper and Bronze Pigments

Copper and bronze pigments are either pure copper (copper color) or an
alloy of copper and zinc (golden bronze color). As the percentage of zinc
in the alloy increases, the shade of the bronze shifts greener. These pigments are produced in the same type of ball mill process that is employed in
making the standard cornflake grade aluminum pigments, which involves
milling the pigment with round balls in a horizontal cylinder with solvent.
Figure 10.14 shows an SEM photograph of a typical 35 μm cornflake structure bronze pigment. Table 10.2 shows the typical composition of the various
pigments that are commercially available today.

Figure 10.14   35 μm bronze pigment.
Source: Courtesy of Exkart.

Table 10.2   Copper and bronze pigment composition.

Color	% Copper	% Zinc
Copper	100%	0%
Pale gold	90%	10%
Rich pale gold	85%	15%
Rich gold	70%	30%

Table 10.3   Copper and bronze pigment regulations.

Area of use	United States	European Union	China	Japan
Mucous membrane – oral	Permitted	Permitted	Permitted	Not regulated
Eye area	Permitted	Permitted	Permitted	Not regulated
External	Permitted	Permitted	Permitted	Not regulated
Rinse off	Not applicable	Permitted	Permitted	Not regulated

Copper and Bronze Pigment Regulations

Table 10.3 shows the regulatory status of copper and bronze pigments in the major world markets. As can be seen, these pigments are regulated in the United States, the European Union, and China. They are also not regulated in Japan, because they do not fall into the category of Coal Tar Colorants.

The major difference between aluminum and copper/bronze pigments is that the latter are permitted for ingested use in the United States, making them available for formulation in a much wider range of decorative cosmetics. Like the aluminum pigments, there is a particle size restriction in that market, which is 45 μm. In Europe, copper is listed in Annex IV of the EU Cosmetic Regulation (European Commission 2009) with no restrictions. As it is not a food colorant, there are no specifications, so the bronze is considered an extension of copper. The same situation applies in China.

Using Copper and Bronze Pigments

Copper and bronze pigments are widely used in most types of decorative cosmetics, including lip, nail, and eye products. Unlike aluminum, they do not react with water, so there is no issue with hydrogen generation, but they can tarnish in aqueous systems, limiting their use therein. Copper and bronze pigments can also react with nitrocellulose in nail polish, which sometimes results in a slight gelling of the finished product, so caution must be used in these systems. To avoid these issues in both aqueous and nail polish formulations, encapsulated grades are available, which render the pigments inert.

Based on their thin structures, copper and bronze pigments are susceptible to deformation when exposed to high shear forces, so they must not be milled in any way. The reader is referred back to Figure 10.13 to see how harsh milling can affect metallic pigments.

Fluorescent and Phosphorescent Pigments

There are a number of terms ending in "-escence" that describe various light effects. Before moving headlong into a review of fluorescent and phosphorescent pigments, these need to be defined in order to clarify the differences among them.

Incandescence: This is light emitted from a substance as a result of subjecting said substance to heat. Think of an incandescent light bulb – don't touch it!

Luminescence: This is light emitted by a substance that does not result from the heating of that substance. It may be due to chemical reaction, electrical energy, subatomic motions, or stress on a crystal.

Fluorescence: This is light immediately emitted by a substance that is of a different wavelength than the light absorbed by that substance. Generally, the emitted light is of a longer wavelength than the absorbed light. The most spectacular effect of fluorescence occurs when the absorbed light is in the ultraviolet region, so is invisible to the human eye, and the emitted light is in the visible region, so is observable by the human eye. A common example in daily life is a fluorescent light bulb.

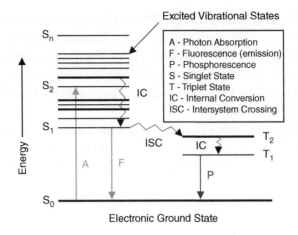

Figure 10.15   Jablonski diagram.
Source: Courtesy of Day-Glo.

Phosphorescence: This is light emitted in a different wavelength from absorbed light, but where the light is not emitted immediately, but rather the emission stretches over several hours. Phosphorescent colors are commonly referred to as "glow in the dark."

Chemiluminescence: This is light emitted as a result of a chemical reaction with little or no heat emission.

Bioluminescence: This is light produced from a living organism through a biochemical reaction. The lightning bug (firefly) is a great example.

The mechanism for all these light emitting effects is well known and is expressed in the Jablonski diagram, shown in Figure 10.15. Arrow A represents light absorption from the ultraviolet region, while arrows F and P represent light being subsequently emitted in the visible region.

All of the light emitting pigments used in decorative cosmetics are of the type where the absorbed light is in the ultraviolet range and the emitted light is in the visible range. The majority of these light emitting pigments are fluorescent, but there is one that falls into the phosphorescent category. Since they are the most voluminous in terms of sheer number, the fluorescent pigments will be reviewed first.

Fluorescent Pigments

Fluorescent dyes date back, in many cases, to the late nineteenth century. As discussed in Chapter 1, dyes are not very useful in decorative cosmetics because of their solubility and lack of opacity. Day-Glo Color Corporation

(Cleveland, OH, USA) solved this problem by incorporating US CFR Title 21, Parts 1–99 listed dyes into resin matrices, resulting in pigments that perform well in decorative cosmetic products. As of the publication of this book, Day-Glo is the sole producer of cosmetic grade fluorescent pigments globally, so it is this company's methodology and products that will bear the focus of a large part of this section.

These fluorescent pigments, a range of which are shown in Figure 10.16, are produced by incorporating a US Food and Drug Administration (FDA) certified primary color (dye) into a resin matrix at elevated temperature,

Figure 10.16   (a) Violet fluorescent pigment. (b) Pink fluorescent pigment. (c) Blue fluorescent pigment. (d) Orange fluorescent pigment. (e) Yellow fluorescent pigment.
Source: Courtesy of Day-Glo.

followed by cooling and subsequent particle size reduction. On the surface, this process seems very simple, but there are many nuances to it that make consistent quality pigments difficult to manufacture. The final pigments are considered preparations similar to the absorption pigment dispersions discussed in Chapter 6. The dyes are uniformly distributed throughout the resin matrix and are not chemically altered or chemically combined with the resins.

Day-Glo's commercially available cosmetic-grade fluorescent pigments are designated as Elara™ series pigments. They are designed for use in oil, solvent, and water based cosmetic formulations. These colorants employ a thermoset polymer that is resistant to the solvents common in nail enamels and hair sprays. They perform well in all types of personal care products, including lip gloss and lip sticks, lotions and soaps, face paint, temporary hair color, and nail lacquer. Some of the products are single-colorant products, while others have multiple colorants in them. Table 10.4 shows the products in the range, along with their compositions and color space. The resin employed in these products carries the INCI name: Norbornanediamine/Resorcinol Diglycidyl Ether Crosspolymer.

Fluorescent Pigment Regulations

Table 10.5 shows the regulatory status of the Elara™ series fluorescent pigments in the United States, European Union, and Japan. They are regulated in these markets based on the colorants used to make them, while the base resin is considered as a cosmetic ingredient, and thus is not regulated specifically This means that the formulator must follow the regulations for the absorption color used to make the fluorescent pigment in question. For example, Elara™ EL-11, listed in Table 10.5, is made using D&C Red No. 28

Table 10.4   Elara™ series pigments.

Product code	Color	Composition
EL-11	Aurora Pink	D&C Red No. 28 D&C Red No. 22
EL-13	Rocket Red	D&C Orange No. 5 D&C Red No. 28 D&CC Red No. 22
EL-15	Blaze Orange	D&C Yellow No. 10 D&C Red No. 28 D&C Red No. 22
EL-17[a]	Saturn Yellow	D&C Yellow No. 10
EL-20	Ultra Violet	External D&C Violet No. 2 D&C Red No. 28
EL-21	Corona Magenta	D&C Red No. 28 D&C Red No. 22
EL-37[a]	Horizon Blue	FD&C Blue No. 1

[a]Does not fluoresce under ultraviolet (black light).

Table 10.5 Regulation of the Elara™ series pigments.

Product code	United States	European Union	Japan
EL-11	Permitted[a]	Permitted	Permitted
EL-13	Permitted[a]	Permitted	Permitted
EL-15	Permitted[a]	Permitted	Permitted
EL-17	Permitted[a]	Permitted	Permitted
DGS-20	Permitted[a,b]	Permitted	Permitted
DGS-21	Permitted[a]	Permitted	Permitted
EL-37[c]	Permitted	Permitted	Permitted

[a]Not permitted in the area of the eye.
[b]External only.
[c]Nail and hair only.

and D&C Red No. 22. Neither of these colorants is permitted for use in eye makeup in the United States, so the fluorescent pigment made from them is also not permitted for this use in the United States. It is the formulator's responsibility to carefully check the regulatory status of any colorants used in all countries in which the product he/she is developing are intended to be sold.

Although China follows EU regulations with respect to color additives, as of this writing, the Elara™ pigments are not permitted in China, pending approval for cosmetic use of Norbornanediamine/Resorcinol Diglycidyl Ether Crosspolymer, the base dispersing resin. As mentioned in Chapter 2, prior to being used in cosmetics, a raw material not already listed in the Inventory of Existing Cosmetic Ingredients in China (IECIC) must be submitted for approval to the China National Medical Products Administration (NMPA).

Using Fluorescent Pigments

Bearing in mind the regulatory considerations, the Day-Glo fluorescent pigments are used in a wide variety of decorative cosmetic products. The Elara™ series finds use in lipstick, lip gloss, and eye products (except in the United States), as well as in nail polish. Due to their nature, these encapsulated fluorescent pigments do not require the degree of dispersion that is required of the standard absorption pigments. They will basically stir into the formations that they are intended to color.

Fluorescent pigments minimize some of the problems associated with using the neat colorants from which they are made. The two main areas of improvement are light stability and bleeding/staining. Furthermore, because of the cross-linked nature of these materials, formulations can be

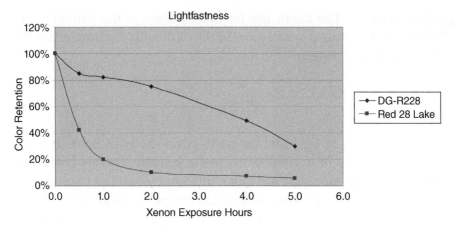

Figure 10.17   Light stability comparison.
Source: Courtesy of Day-Glo.

processed at higher temperatures without adverse effects or migration of color. The fluorescein dyes used in the manufacture of fluorescent pigments are extremely fugitive to light. This instability is somewhat improved when they are incorporated into the resin matrix. However, caution must still be exercised when using the fluorescent pigments in products packaged in transparent containers. The extent of the improvement obtained is demonstrated in Figure 10.17, showing the light stability of D&C Red No. 28 as a neat colorant (line marked with squares) versus the encapsulated fluorescent pigment made from it (line marked with diamonds).

Economics of Fluorescent Pigments

As with all other pigments, market prices vary dramatically based on country of origin, transportation costs, duties, and currency fluctuations (see Chapter 5's discussion on economics for further insight). With these factors in mind, average US market prices for the Elara™ fluorescent pigments currently range from $140 to $155/kg.

Phosphorescent Pigments

There is only one phosphorescent pigment that is permitted for use in decorative cosmetic products in the United States, and that is zinc sulfide. From a regulatory perspective, it is permitted in the United States for use in face makeup formulated for occasional use and for everyday use in nail polish. It is not permitted in either the European Union or China, and it is not regulated in Japan. The major use for this colorant in the United States is in Halloween makeup, as it will glow in the dark with a yellow-green phosphorescence of wavelength ~530 nm.

Glitter

Glitter is unlike any of the other pigments covered in this book. It is insoluble in the majority of decorative cosmetic systems, and its composition and manufacturing processes are quite different from other pigment types. Glitter's structure falls into one of two categories. Traditionally, it is a single sheet of copolymer plastic on to which very thin layers of aluminum and an absorption pigment are deposited. The newer generation of glitters, however, are produced by laminating together two very thin layers of polymers with different indices of refraction, which results in light interference, causing the glitter to exhibit both a reflected and a transmitted color. The color will vary based on the thickness of the laminated particle. The principle with which color is produced by this laminated material is the same as that described in Chapter 9 for the interference (iridescent) pearl pigments. The typical polymers used in this type of glitter are polyethylene terephthalate, acrylate copolymer, and polymethyl methacrylate. Examples of silver, iridescent, and multicolored glitters are shown in Figure 10.18.

Once the sheets are produced by either mechanism, they are cut into small hexagonal pieces ranging from 0.002″ to 0.032″ in size, or into square pieces ranging from 0.004″ to 0.040″.

American Glitter Inc., a major global producer of cosmetic glitters, markets a typical range of products in both the conventional and the lamination categories. These will be used as examples to further our discussion. Both varieties are available in hexagonal and square formations in the sizes shown in Table 10.6.

The color palette available for the traditional types of glitter products is shown in Table 10.7, while Table 10.8 shows the color range of the laminated glitters.

Glitter Regulations

There are no specific regulations covering glitter in any of the major world markets. In spite of this, they are generally considered compliant due to the fact that the traditional cosmetic grades consist of physical combinations

Figure 10.18 (a) Silver glitter pigment. (b) Iridescent glitter pigment. (c) Multi-color glitter pigment.
Source: Courtesy of American Glitter.

Table 10.6 Glitter particle size ranges.

Shape	Size 1	Size 2	Size 3	Size 4	Size 5	Size 6
Hexagonal[a]	0.002″	0.004″	0.008″	0.015″	0.025″	0.032″
Square[b]	0.004 × 0.004″	0.008 × 0.008″	0.015 × 0.015″	0.025 × 0.025″	0.032 × 0.032″	0.040 × 0.040″

[a]Point-to-point measurement.
[b]Edge measurement.

of polymers that are well known and considered safe, with absorption colors that are found on the positive lists in the major world markets. However, there is one caution with respect to this: the limiting factor concerning traditional glitter use is the regulatory restrictions placed on the absorption color used in the market where the finished cosmetic will be sold. For example, seven of the American Glitters products are made with ferric ammonium ferrocyanide, which is not permitted for use in the United States in cosmetics that can come in contact with the mucous membranes, including lips, so these particular glitter products cannot be applied in this fashion within

Table 10.7   Cosmetic glitters (traditional).

Product code	Polymer	Aluminum	Colorant
Water Clear	PET	None	None
Real Silver 0011	PET	Yes	None
Antique Silver 0012	PET	Yes	D&C Red No. 7 Ca Lake
Brazilla Green 0014	PET	Yes	Ferric ammonium ferrocyanide + FD&C Yellow No. 5 Al Lake
Rocky Green	PET	Yes	Ferric ammonium ferrocyanide + FD&C Yellow No. 5 Al Lake
All Blue 0015	PET	Yes	Ferric ammonium ferrocyanide + D&C Red No. 34 Ca Lake
Persian Blue 0016	PET	Yes	Ferric ammonium ferrocyanide
Ojos Emerald 0017	PET	Yes	Ferric ammonium ferrocyanide + FD&C Yellow No. 5 Al Lake
Night Lavender 0018	PET	Yes	Ferric ammonium ferrocyanide + D&C Red No. 34 Ca Lake
Indian Purple 0019	PET	Yes	Ferric ammonium ferrocyanide + D&C Red No. 34 Ca Lake
Meccana's Gold 0020	PET	Yes	D&C Red No. 7 Ca Lake + FD&C Yellow No. 5 Al Lake
Mint Bronze 0021	PET	Yes	D&C Red No. 34 Ca Lake + FD&C Yellow No. 6 Al Lake

PET, polyethylene terephthalate.

Table 10.8   Laminated cosmetic glitters (crystal).

Product code	Polymer 1
Ice Crystal 0030	PET
Holo Silver 0901	PET
Iridescent 4801	PET
Iridescent 4802	PET
Iridescent 4803	PET
Iridescent 4804	PET

PET, polyethylene terephthalate.

this market. As always, the cosmetic formulator must have regulations in the forefront of his or her mind.

The polymers that are used in American Glitters proprietary Cosmetic Glitters and Cosmetic Crystals lines all have INCI names: Polyethylene Terephthalate, Acrylate Copolymer, and Polymethyl Methacrylate.

Using Glitter

Glitters can be used in many types of decorative cosmetics to add "pop" to the products. They are particularly common – regulations permitting – in eye makeup and for general body decoration. The formulator's choice of glitter size, shape, and color is strictly based on the effect desired in the finished cosmetic. Typically, the smaller-sized glitters are used in eye shadow and blush so as not to make the product effect too brash. The larger-sized ones are used more in nail polish, where the square or hexagonal shapes enhance the color of the polish. Glitters, in general, find use predominantly in makeup targeted to children and teenagers. Not many grandmothers care for the "glittery" look of these products.

Due to incompatibility between some of the polymers and the solvents used in nail polish, the formulator must carefully choose a glitter that will be compatible with the other ingredients in the system.

Economics of Glitter

As with all other pigments, market prices for glitters in different countries vary dramatically based on country of origin, transportation costs, duties, and currency fluctuations. The following are some average market prices for glitters, which can of course be affected by the aforementioned factors:

Traditional Glitters: $65–120/kg.
Laminated Glitters: $105–190/kg.

Silver and Gold Pigments

Silver and gold pigments are essentially pure precious metals. They are similar to the aluminum, copper, and bronze pigments in that their luster and sparkle increase with particle size, while their coverage decreases. However, the brightness of the effect obtained with gold and silver pigments is much more dramatic than that of those composed of aluminum, bronze, or copper.

The EU purity requirements (European Commission 2009) for gold permit silver and copper to be present in it up to 7% and 4%, respectively. This allowance, coupled with the fact that it is available in several particle size ranges, results in a variety of shades and effects for gold pigment. With global gold metal prices in the range of $60/g ($60 000/kg) at the time of this book's writing, there is not much demand for gold as a cosmetic colorant. Because of this, the only world producer of gold

Table 10.9   Gold pigment products.

Product code	Gold	Silver	Copper
Gold 300	x	x	
Gold 834	x	x	
Gold 889	x	x	x
Gold 955	x	x	
Gold 975	x		
Gold 985	x	x	

Table 10.10   Gold pigment particle sizes.

Form	Particle size number	Particle size range
Powder	No. 7	$<600\,\mu m$
Powder	No. 8	$<425\,\mu m$
Powder	No. 9	$<400\,\mu m$
Flakes	No. 5	$<2.5\,\mu m$
Flakes	No. 6	$<1.0\,\mu m$

pigments for cosmetics is J.G. Eytzinger GmbH. The company's products, by virtue of their exclusivity, are our sole usable examples of commercially available material. Their range of products, showing the different combinations of the metals, are listed in Table 10.9, while Table 10.10 outlines the particle size ranges available for different effects and application forms.

In addition to gold pigment, Eytzinger supplies a cosmetic grade silver pigment, as does Linbraze S.r.l. which offers food and cosmetic grade silver pigments that are not merely variations of "whites." Although far less expensive than gold, the global price of silver metal is in the range of $560/kg – still a high price for a starting material for pigment production.

Silver and Gold Pigment Regulations

Table 10.11 shows the regulatory situation in the four major world markets. As previously mentioned, the regulation in the European Union for gold allows for up to 7% silver and 4% copper, provided the gold content is a minimum of 90% (European Commission 2009). EU regulations for silver require purity of a minimum of 99.5% (European Commission 2009), while the US requirement is a minimum of 99.9% (FDA 2018).

Table 10.11   Regulations for gold and silver pigments.

Product	United States	European Union	China	Japan
Silver	Permitted[a,b]	Permitted[c]	Permitted	Not regulated
Gold	Not permitted	Permitted[d]	Permitted	Not regulated

[a]Permitted for nail polish only; maximum level 1%.
[b]Purity must be ≥99.9%.
[c]Purity must be ≥99.5%.
[d]Purity must be ≥90.0%.

Economics of Silver and Gold Pigments

There is often uncertainty in the global economy, which causes volatility in the precious metals market. Therefore, it is not possible to list market pricing beyond the estimates already given for the gold and silver pigments. Suffice it to say that these are the most expensive pigments used in the manufacture of decorative cosmetics.

Summary

Today's user market for decorative cosmetic products is consumed by the individual's need to make appearance a personal statement. The traditional absorption and pearl pigments, whilst essential elements of the color palette, do not meet all of the requirements of this insatiable desire for spectacular effects. The colorants reviewed in this chapter do meet these needs, and therefore now hold an important place on the formulator's palette. Their importance is expected to grow as the trend for special effects in cosmetics grows across the globe. In the next chapter, another competing growing class of colors, the natural colors, will be covered.

References

European Comission (2009). Regulation (EC) no. 1223/2009. *Official Journal of the European Communities.* https://eur-lex.europa.eu/legal-content/EN/ALL/?uri=CELEX%3A32009R1223 (accessed September 28, 2020).

FDA (2018). Code of Federal Regulations, Title 21 Part 73.500, revised as of April 1, 2018. https://www.accessdata.fda.gov/scripts/cdrh/cfdocs/cfcfr/CFRSearch.cfm?CFRPart=1-91 (accessed September 28, 2020).

NMPA. (2015). Safety and Technical Standards for Cosmetics. National Medical Product Administration, People's Republic of China, Beijing, China (2015 edition) http://www.sesec.eu/app/uploads/2016/02/Cosmetics-Safety-and-Technical-Standards-2015-Version-Foreword-and-summary.pdf (accessed September 28, 2020).

Chapter 11

Natural Colorants

In today's world of decorative cosmetics, the word "natural" evokes a sense of excitement and well-being from consumers and cosmetic industry insiders alike. There are many conferences held, papers presented, and articles written that focus on natural cosmetics and their ingredients. This has led to many new companies specializing in the marketing of natural cosmetic products and numerous traditional cosmetic companies launching products that fall into the naturals category.

A number of natural colorants are included in the positive lists of approved colorants for the manufacture of cosmetic products of the United States, the European Union, and China, which would appear to make possible the development of "natural" cosmetics. However, the vast majority of these leave a lot to be desired with respect to their color properties. They tend to lack intensity, exhibit rather dull shades, and suffer from a number of stability issues, all of which limit their use in color cosmetics. Yet, even with these issues, the importance of natural cosmetics in today's global environmentally conscious, "green" focused populace, without whose purchasing power the question of cosmetics formulation would be moot, justifies a review of these colorants.

Coloring the Cosmetic World: Using Pigments in Decorative Cosmetic Formulations,
Second Edition. Edwin B. Faulkner. Edited by Jane C. Hollenberg.
© 2021 John Wiley & Sons Ltd. Published 2021 by John Wiley & Sons Ltd.

Definition of "Natural"

There are many definitions of the term "natural," and more of them are constantly being created. Why? Mostly because the criteria tend to be fluid, which is to say that the term "natural" is not subject to the same stringent restrictions of use as a term such as "organic," which is often uttered in the same breath. To stimulate the reader's appetite, a few definitions and criteria excerpted from the Natural Ingredient Resource Center's (NIRC) website (www.naturalingredient.org) follow.

Natural Ingredient Criteria

Natural ingredients include plant, animal, mineral, or microbial ingredients that are:

- present in or produced by nature.
- produced using minimal physical processing.[1]
- directly extracted using simple methods, simple chemical reactions, or naturally occurring biological processes.[2]

Natural ingredients are ...

- grown, harvested, raised, and processed in an **ecological** manner.
- **not** produced synthetically.
- free of all petrochemicals.
- **not** extracted or processed using petrochemicals.
- **not** extracted or processed using anything other than natural ingredients as solvents.
- **not** exposed to irradiation.
- **not** genetically engineered & do not contain GMOs (genetically modified organisms).

Natural ingredients do ...

- **not contain** synthetic ingredients.[3]
- **not contain** artificial ingredients including colors or flavoring.
- **not contain** synthetic chemical preservatives.

---

[1] Minimal Processing means the ingredient has had no more processing than something which could be made in a household kitchen, stillroom, on a farm, or vineyard. It doesn't mean they have to actually be made in those settings, but that they would require no more equipment or technology than that which could be employed in those settings.

[2] Simple Extraction Methods/Simple Chemical Reactions include cleaning, cold pressing, dehydration, desiccation, drying, evaporation, filtering, grinding, infusing [water or natural alcohol], & steam or water distilling.

[3] Produced by synthesis, a compound made artificially by chemical reactions, from simpler compounds or elements. The NIRC has allowed for an exception in the case of "lye" in the manufacture of soap.

The United States Food and Drug Administration (FDA) has no official definition of "natural," but generally refers to natural ingredients as being extracted directly from plants or animal products as opposed to being produced synthetically. The FDA also considers that "natural ingredients are derived from natural sources" (e.g., soybeans and corn provide lecithin to maintain product consistency; beets provide beet powder used as food coloring). Other ingredients are not found in nature and therefore must be synthetically produced as artificial ingredients. Also, some ingredients found in nature can be manufactured artificially and produced more economically, with greater purity and more consistent quality, than their natural counterparts. For example, vitamin C/ascorbic acid may be derived from an orange or produced in a laboratory. Food ingredients are subject to the same strict safety standards regardless of whether they are naturally or artificially derived (FDA 2004).

With these elements of the natural world in mind, natural color additives for cosmetics can now be considered.

Natural Color Regulations

Natural colorants are regulated in the United States, the European Union, and China; they are not regulated in Japan, as only synthetic coal tar colors are regulated there. In the United States, all natural colors are covered in Code of Federal Regulations Title 21, Part 73, whereas the synthetic organic colors are covered as certifiable colors in CFR Title 21, Part 74 (FDA 2018a). In the European Union, all colorants, natural and synthetic, are covered in Annex IV of the Cosmetics Directive (European Union 2009), and in China, natural colors are covered in Table 6 of the Safety and Technical Standards for Cosmetics (2015 version) (NMPA 2015).

Table 11.1 shows the common natural colors that are approved in the United States, alongside their status in the other major world markets. What follows is a consideration of each in turn.

Annatto

Table 11.1 Color regulations for natural colorants approved in the United States.

Colorant	United States	European Union	China	Japan	Comments
Annatto	21CFR73.2030	CI 75120 E 160(b)	All products	Not regulated	E160(ii) E number not listed in Annex IV
B-Carotene	21CFR73.2095	CI 40800	All products	Not regulated	
Caramel	21CFR73.2085	Caramel E 150a	All products	Not regulated	
Carmine	21CFR73.2087	CI 75470 E120	All products	Not regulated	
Chlorophyllin Cu Complex	21CFR73.2125 Use in dentifrices that are drugs	75 810 (Annex IV)	All products	Not regulated	E140/141(ii) listed as CI 75815
Guaiazulene	21CFR73.2180 External only	Not permitted	Not permitted	Not regulated	
Henna	21CFR73.2190 Hair only	Permitted	Not permitted	Not regulated	
Guanine	21CFR73.2329	Not permitted	Not permitted	Not regulated	

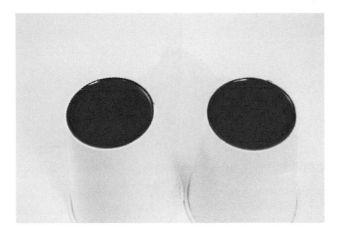

Figure 11.1   β-Carotene (l) and Annatto (r).
Source: Courtesy of Sensient Color Technologies.

Annatto (shown in a liquid form alongside β-Carotene in Figure 11.1) is a yellow to orange colorant that has seen use for hundreds of years, particularly in Central and South America, where its uses have run the gamut from food to ritual face painting to textile coloration. It is even believed to repel insects and to treat colds and fevers. Today, it is most commonly used to color cheese, baked goods, and snack foods.

Annatto originates from the evergreen shrub *Bixa oreallana*, which grows in Latin America, the Caribbean Islands, India, and parts of Africa. It is available in an alkali-soluble form called Norbixin and an insoluble form called Bixin. It is not stable to prolonged direct exposure to light or acid media, where it is susceptible to chemical changes.

Annatto is approved for use in all cosmetic products in the United States, the European Union, and China.

Chlorophyllin Copper Complex

Figure 11.2   Chlorophyllin-Copper Complex.
Source: Courtesy of Sensient Color Technologies.

Chlorophyllin-Copper Complex is a green colorant (Figure 11.2) produced from chlorophyllin extracted from alfalfa by saponification, the magnesium of which is replaced with copper. It is available in two forms, one which is water soluble (the sodium salt) and one which is not. Chlorophyllin-Copper Complex is stable to light and to alkali, but the sodium salt will precipitate in acid solutions.

Chlorophyllin-Copper Complex is limited to coloring cosmetic dentifrices in the United States.

Carmine

Carmine is a bright blue shade red colorant which is the aluminum lake of carminic acid, the processed extract of dried female cochineal beetles (see Figure 3.17). The beetles are harvested predominately in Peru, though some come from the Canary Islands. It takes approximately 155,000 insects to make one kilogram of carmine.

This colorant was used for centuries in the coloring of textiles before the birth of the synthetic dyestuff industry in the mid- to late nineteenth century. Its most famous use was as the coloring matter for eighteenth and nineteenth century red British army coats, leading to British troops being known as "Red Coats." The main use of carmine today is in the food industry, to color the Italian aperitif, Campari.

Carmine stands apart from the other natural colorants because it is a strong, vibrant, bright red shade. For a natural colorant, its light stability is acceptable in dry form but poor in aqueous suspension. Carmine is not stable in acid systems or at temperatures that exceed $60\,^\circ$C. Also, being a chemical lake, it will break down in strong acids and bases, liberating free carminic acids.

Carmine is approved for all cosmetic uses in the United States, the European Union, and China. The main reason for carmine's continued use in color cosmetics is that it is the only clean, bright red pigment permitted for eye area use in the United States. Like other non-coal tar colors, it is not regulated in Japan.

Henna

Henna is a dye extracted from the leaves of the Henna plant, most of which come from India. It is predominantly used in the coloring of hair, for a conditioning as well as a coloring effect. Henna is also used as a temporary skin dye in various religious ceremonies and decoratively in some Middle Eastern and Asian cultures.

In the United States, henna is limited to coloring of the hair only; use in the coloring of eyelashes and eyebrows is specifically prohibited because of concern that it could find its way into the eye, causing irritation.

β-Carotene

β-Carotene (Figure 11.1) is a yellow to orange pigment that is obtained from the fungus *Blakeslea trispora*. It has good stability to light, heat, acids, and alkalis.

In the United States, it is approved for all types of cosmetic products.

Caramel

Caramel (Figure 11.3) is a brown colorant that is produced by the burning of any one of a number of sugars, including dextrose, invert sugar, lactose, malt syrup, molasses, starch hydrolysates, and sucrose (FDA 2018b). It has excellent light and heat stability and is stable in both acid and alkaline systems. Caramel is available in both powder and liquid form, the former being used in anhydrous systems such as eye shadow, blush, and lipstick, the latter in aqueous systems or in the water phase of emulsion products.

Caramel is approved for all types of cosmetic uses in the United States.

Guaiazulene

Guaiazulene is a derivative of azulene, a main component of chamomile, which has been used for generations to treat many ailments, from insomnia to minor wounds to ulcers. It is a blue colorant (*azul* is the Spanish word for blue).

Figure 11.3 Caramel.
Source: Courtesy of Sensient Color Technologies.

Guaiazulene is approved for external use only in the United States and is not approved for any cosmetic use in the European Union or China. This limited approval, plus the fact that guaiazulene sells for over $220/kg, restricts use.

Guanine

Natural pearl essence, also known by the chemical name guanine was covered in Chapter 9. Due to high cost and limited availability, it is currently seldom used.

Guanine is approved for all types of cosmetic products in the United States, but is not permitted in either the European Union or China. There has not been any interest in the cost and effort involved in obtaining approval in either of these markets.

Natural Colors Permitted for Cosmetic Use Outside the United States

There are some natural colors that are not permitted for use in the United States but are permitted in the European Union and China (and not regulated in Japan). Table 11.2 shows their regulatory status. Discussion of the individual colors follows.

Because the trend toward using natural ingredients in decorative cosmetics is fairly recent, there is scant data on the use of these natural colorants in color cosmetics. The formulator is encouraged to experiment with them in an effort to determine their value in natural color cosmetics.

Lycopene

Table 11.2   Color regulations for natural colorants not approved in the United States.

Colorant	United States	European Union	China	Japan	Comments
Lycopene	Not permitted	CI 75125 E160(d)	All products	Not regulated	
Vegetable Carbon	Not permitted	CI 77266 E152	All products	Not regulated	
Curcumin	Not permitted	CI 75300 E100	All products	Not regulated	
Capsanthin/ Capsorubin	Not permitted	E160(c)	All products	Not regulated	Common name Paprika; no CI number
Beta-Apo-8'- Carotenal	Not permitted	CI 40820 E160(e)	All products	Not regulated	
Canthaxanthin	Not permitted	CI 40850 E161(g)	All products	Not regulated	

Figure 11.4   Lycopene.
Source: Courtesy of Sensient Color Technologies.

Lycopene is chemically a carotenoid, similar in structure to β-Carotene (Figure 11.4). It is a water-insoluble pigment that is extracted from red tomatoes. Its shade is in the yellow-red range, not the typical intense red of tomatoes. It can be used to color many types of decorative cosmetics.

Figure 11.5   Curcumin.
Source: Courtesy of Sensient Color Technologies.

Curcumin

Curcumin (Figure 11.5) is a yellow pigment that is responsible for the yellow color of the spice turmeric, from which it is extracted. In decorative cosmetics, it is most useful in eye makeup, where yellow colors have the most value.

Vegetable Carbon

Vegetable Carbon is cousin to D&C Black No. 2. The difference between them is that Vegetable Carbon is made from the carbonization of vegetable matter, while D&C Black No. 2 is made from oil via the furnace process. Typical vegetable matter used in the manufacture of Vegetable Carbon includes wood, cellulose residues, and the shells of nuts. As Vegetable Carbon is pure carbon, and thus chemically identical to D&C Black No. 2, it has the same physical and color properties. The most common uses are in eye area products such as mascara and eyeliner, where it provides a very intense and useful black shade.

Capsanthin/Capsorubin

Capsaicin

Capsanthin

Capsorubin

These two pigments are extracts from the sweet red pepper and are together commonly known as paprika. They are extracted together with such solvents as methanol, ethanol, and acetone, and operate in tandem to produce an orange color, which is used most effectively in eye makeup. There is no CI number for these materials.

The other component of the extract is capsaicin, which is the chemical that accounts for the spice flavor of paprika. Grades from which the capsaicin has been removed in order to reduce the potential for skin and eye irritation are preferable for cosmetic use.

Beta-Apo-8′-Carotenal

Beta-Apo-8′-Carotenal is an orange colorant that is a member of the carotenoid family, along with β-Carotene. It can be extracted from various natural sources such as oranges or can be synthetically produced.

Beta-Apo-8′-Carotenal is available in both water- and oil-dispersible forms and is best used in eye makeup and blush.

Canthaxanthin

Canthaxanthin is another member of the carotenoid family, so like the others covered in this chapter, it exhibits a yellow-red shade. It is found naturally in mushrooms and in some species of shrimp, but is also available synthetically. Like the other colorants in its class, it is most useful in eye shadow and blush.

Other Natural Colors

There are quite a number of other natural colors available, but they are not useful for decorative cosmetics, either because they are dyes, or because they are not on the positive lists in any of the major markets (or both).

The reader has undoubtedly noticed that "where used" data and other data on colors covered in this chapter are significantly less complete than for the pigments discussed in the previous chapters. This is because there was little interest in the past in using these colorants due to their deficiencies in stability and tinting strength in comparison to their synthetic counterparts. With the current emphasis on natural ingredients and cosmetics, this situation can be expected to change in the future.

References

European Union. (2009). Regulation (EC) No. 1223/2009 of the European Parliament and of the Council of 30 Nov 2009. *Official Journal of the European Communities.* http://eur-lex.europa.eu/LexUriServ/LexUriServ.do?uri=OJ:L:2009:342:0059:0209: en:PDF (accessed September 28, 2020).
FDA. (2004). Food Ingredients and Colors, International Food Information Council (IFIC) and US Food and Drug Administration, November 2004; revised April 2010.

`https://www.fda.gov/food/food-ingredients-packaging/overview-food-ingredients-additives-colors` (accessed September 28, 2020).

FDA. (2018a). Code of Federal Regulations Title 21, Part 73, revised as of April 1, 2018. `https://www.accessdata.fda.gov/scripts/cdrh/cfdocs/cfcfr/CFRSearch.cfm?CFRPart=73` (accessed September 28, 2020).

FDA. (2018b). Code of Federal Regulations Title 21, Part 73.85, revised as of April 1, 2018. `https://www.accessdata.fda.gov/scripts/cdrh/cfdocs/cfcfr/CFRSearch.cfm?fr=73.85` (accessed September 28, 2020).

NMPA. (2015). Safety and Technical Standards for Cosmetics. National Medical Product Administration, People's Republic of China, Beijing, China (2015 edition) `http://www.sesec.eu/app/uploads/2016/02/Cosmetics-Safety-and-Technical-Standards-2015-Version-Foreword-and-summary.pdf` (accessed September 28, 2020).

Chapter 12
Some Slices of Life

Editor's Note: Ed Faulkner's observations are still instructive nine years after he first made them. I have only changed some time references and added a few current comments of my own.

If you've gotten this far in the book, congratulations, you now have enough information to be dangerous! The next step in your journey along the path to color expertise is to take the information that you have absorbed from this book and put it to practical use in the laboratory or cosmetics factory where you spend your days. I hope that you have found the topics covered valuable.

I have seen tremendous changes in the pigment and cosmetics industries in my 40 years in the wilderness of life (Moses didn't have anything on me!). In the arena of cosmetic pigments, with the exception of my home for over 40 years, Sun Chemical, which retains its identity within the DIC Group, most of what were, years ago, the major players in the market – such as H. Kohnstamm, Clark Colors, Thomasett Colors, Williams-Hunslow, Mearl, Max Marx, les couleurants Wackherr, Rona Pearl, and Hilton Davis – have been absorbed by other companies; companies we know well today by the names Sensient Technologies, Dystar, DIC, and Merck KGaA.

Pigments were much simpler in the old days, because your only choices were standard certified organic colors, iron oxides, titanium dioxide, inorganic pigments, bismuth oxychloride, and titanium dioxide-coated micas. In today's world of color, you have a myriad of choices beyond those, including surface-treated pigments, dispersions, fluorescent pigments, glitter, metallic pigments (aluminum, copper, bronze), and a host of effect pigments on

Coloring the Cosmetic World: Using Pigments in Decorative Cosmetic Formulations,
Second Edition. Edwin B. Faulkner. Edited by Jane C. Hollenberg.
© 2021 John Wiley & Sons Ltd. Published 2021 by John Wiley & Sons Ltd.

various substrates from synthetic mica to borosilicate to aluminum. Today's formulator can create effects in makeup that were unknown to the person standing in his or her place 50 years ago!

There have likewise been huge changes in the demographics of the cosmetics industry over the years. You may not know it, but 50 years ago, it was common for cosmetics companies to be owned by pharmaceutical companies. Pfizer owned Coty. Eli Lilly owned Elizabeth Arden. Schering Plough owned Maybelline and Rimmel. Squibb owned Charles of the Ritz. As the cosmetics industry grew more competitive with the influx of Asian products into the United States and Europe, however, these subsidiaries' profit profiles no longer matched up with those of the pharmaceutical giants, so the companies were divested. A consolidation of the industry has occurred as companies including Estée Lauder, L'Oréal, Shiseido, and Coty have grown enormously through purchases of smaller companies. A more recent development is the success of start-up companies, which have achieved rapid growth through online marketing of products made by private label manufacturers. Among the results today is that research and development is consolidated, with expenditures being significantly lower than they were under the pharma ownership. This has led to cosmetic chemists having much fuller plates today, and thus having only limited time for new raw material approvals.

One of the most interesting phenomena that I have observed is the growth of the cosmetics business in developing countries. As the economies of these countries have started to grow, their citizens have begun to have small amounts of disposable income, and one of the first things their female populations have bought is color cosmetics. These products are relatively inexpensive and provide an almost instant boost to one's self-image. I began visiting such countries long before they became affluent, so I was able to observe the populace without makeup and then see the changes occur over time. I watched the increase in color of the women's faces and hands, and saw the consumption of cosmetic pigments grow in a corresponding manner. Just as the demand for color cosmetics has grown throughout the world, so have companies serving the industry increased in size and number, including raw material suppliers, cosmetic marketers, and private label manufacturers.

Another interesting evolutionary change I have watched is that of the nail lacquer industry in the United States. Little more than 40 years ago, two name-brand cosmetic companies made nail polish in the United States: Revlon and Chesebrough Ponds. The balance of the nail color sold in the country by the US-based companies was made by contract manufacturers. There were a number of privately held companies in this specialty business, including Maas & Waldstein, Kirker Chemical, TEVCO Inc., Glosstex Industries, Decorative Industries, and Chemcoat, all located within 50 miles of New York City. Beginning in the mid 1980s, consolidation began as both Revlon and Chesebrough Ponds stopped production, leaving all of the nail polish manufacture in the hands of the private label producers. Subsequently, the number of these manufacturers was reduced through a series of acquisitions over 20 years, culminating in 2007 with Kirker's purchase of TEVCO, leaving it the only domestic US producer of nail color. The vacuum created by the Kirker purchases was partially filled by the French producer, Fiabila, which set up production in the United States at the former Glosstex facility in Mine Hill, NJ, leaving just two producers of polish in the United States.

A significant addition to the nail color industry has been the development of gel nail technology, where the film on the nails is formed by a chemical reaction, not the traditional solvent evaporation. One wonders just where or how the nail lacquer industry will evolve next, in the coming decades ...

Over the years, I have traveled to almost 40 countries and probably about 100 cities. It was a fantastic experience to meet new people, learn about their cultures, see sights that most people only see in books or movies, and enjoy fabulous food. Throughout the years and the travels involved, I learned a lot of life's lessons the hard way, and, if you will indulge me, I would like to share some of them with you here. Who knows, maybe you'll avoid some hard lessons yourself, so take heed. I find it easier to stick to the lessons I've learned if I have a catchy phrase to remember them by, so I'll share with you these little quips; maybe they'll work for you the same way. I won't take credit for the phrases, because I learned them from other people who passed their lessons along to me the way I'm attempting to do likewise for you. They are in no particular order.

Don't let a camel put his nose under the edge of your tent because, if you do, soon you'll have the whole camel in your tent. Be very careful about making exceptions to the rules or deviating from standard practices. Sometimes, making what appears to be an insignificant concession can have far-reaching consequences down the line, so think about the future implications if you are in a position to change the norm.

It's not what you know, it's who you know. Boy, is this one true! It applies in your personal as well as your professional life. It speaks volumes to the importance of establishing and nurturing relationships. It is impossible for any of us to effectively go through our daily lives without the input and help of other people. Don't be an island!

Trees don't grow to the sky! Just ask those people who were heavily invested in the real estate market in 2008. There is a limit to all growth, so it's important to keep this in mind when planning and executing – you need a plan B!

You can change a person's behavior, but you can't change their personality. This is a very important distinction. One's personality is fixed both genetically and by the many experiences of one's formative years, so it cannot readily be altered. However, it is possible to change one's behavior or that of others. Just concentrate on the right thing.

The voice of the customer is not the only way to plan new products. In the first decade of the twentieth century, Henry Ford was asked his opinion regarding customer wants and needs. He replied that if you ask the customer what he wants, he'll tell you he wants a faster horse. So, while the voice of the customer is important, don't use it exclusively or you'll stifle your creativity!

There is nothing so constant as change. It is human nature to resist change; we all like to stay in our comfort zone. However, change is inevitable, so get ahead of it and embrace it. Don't let it drag you along kicking and screaming, or you'll waste a lot of time and energy.

If you're not the lead dog, the view never changes. This is for my Iditarod friends. (Iditarod, incidentally, for those of you not in the know, is likely the world's foremost dog sled race.) You don't have to be the leader in every facet of your life, but when you are, it is quite exhilarating. So, take charge of something in your life and enjoy the view.

A rising tide floats all boats. If things are going well, it's easy to fall into the trap of thinking that your actions are the cause of it. While that may be true, don't count on it 100%. Again, have a plan B.

Do what's right, not what's expedient. The corollary to this is, "Don't take the easy way out." We all lead very busy lives, which means a lot of multitasking, so we are constantly tempted to take the short cut. Before going down that path, give some serious consideration to the long-term impacts of doing so: How will this affect me tomorrow, or six weeks from now? Who else will this affect, and how? This might just get you to change your mind.

Do what needs to be done, not what you like to do. Imagine it is a bright, sunny, warm day and your "to do" list has you cleaning out the basement closet. Most of us would say, "The closet can wait; I'm going to play golf, go to the beach, drive my convertible ... " – insert whatever activity you thoroughly enjoy doing on a summer afternoon. Many times in your work life, you will be presented with an equivalent decision: do what needs to be done, or do a work activity that you enjoy. Don't succumb to what is enjoyable, do what needs to be done. You'll feel better about yourself after completing the task or activity.

It's easier to gain forgiveness than it is to get permission. If there has been one single principle that I have followed in my 40+ year work span, it is this one. Take charge, make a decision, and do something. If you over-step your bounds, someone will point out the error of your ways to you and you'll know what your boundaries are the next time. And you might just break a few in the process!

It's better to make a mistake of commission than one of omission. Do something, even if it turns out to be wrong. There is nothing wrong with making a mistake, as long as you don't make that same one over and over again. Action is much preferable to inaction.

When you're up to your eyeballs in alligators, it's sometimes diffi-cult to remember that your initial objective was to drain the swamp. Someone or something is always getting in the way of you doing your job. While distractions are a way of life, don't let them interfere with your pri-mary tasks for the day.

People do what you inspect, not what you expect. If you tell some-one in your charge how important something is, but never ask them about it, they will not pay much attention to it. This doesn't mean that you should become a micro-manager; it means that you must check in from time to time with people on the things that are really important.

Don't confuse activity with accomplishment. This applies to your own performance as well as that of other people. Simply being busy is not a measure of success; meeting objectives is. Something must come out of the pipe. I once let someone go (a nice way of saying I fired them) because the person rarely met any objectives. After the fact, I had numerous peo-ple remark to me how surprised they were that this person was terminated because they always seemed so busy. Do not fall into this trap!

If it's not broken, don't fix it. A major temptation that most people have when they move into a new job is to immediately start making changes in an effort to make their mark on the organization. While change is inevitable, those changes made after gaining experience and through thoughtful consideration cause much less disruption to the smooth func-tioning of an organization. So, I urge you learn your new job, department, company, before you start changing things. Who knows, you might even find that some things are working well just as they are.

Well, this brings us to end of our relationship. I hope that you found some value in this chapter, and in the book overall. I know that I found value myself in writing it!

Appendix A
Pigment Test Methods
Courtesy of Sun Chemical Corp.

Test 1 – Castor Oil Evaluation

Purpose: To evaluate cosmetic pigments (organic and inorganic) in a castor oil dispersion.

Process: The pigment is dispersed in castor oil between the glass plates of a Hoover Muller mill for a specified number of revolutions or time period at a specified RPM.

Materials:

Castor oil
Bleaching White (80% zinc oxide)

Equipment:

Analytical balance accurate to 0.001 gm
Hoover Muller
2 3/8″ Palette knife
3″ Wide drawdown knife

Coloring the Cosmetic World: Using Pigments in Decorative Cosmetic Formulations, Second Edition. Edwin B. Faulkner. Edited by Jane C. Hollenberg.
© 2021 John Wiley & Sons Ltd. Published 2021 by John Wiley & Sons Ltd.

Black and white display sheets
3″ × 4″ Templates
4″ × 6″ Plastic shields (for drawdown displays)

Procedure – Masstone Preparation:

On an analytical balance, weigh the appropriate amounts of pigment and castor oil.

Pigment	Color (gm)	Castor oil (gm)
Fluorescein dyes	0.700	1.000
Fluorescein Lakes	0.500	1.000
D&C toners	0.700	1.000
D&C true pigments	0.700	1.000
D&C lakes	0.500	1.000
Inorganic colors	0.700	1.000
Iron oxides	0.800	1.000

Transfer the mixture from the step above to the Hoover Muller. Using two weights on the Muller balance arm, grind the mixture for 1×100 revolutions. Scrape the dispersion off the plates with the palette knife and replace it on the bottom plate at a distance approximately one-half of the plate radius from the center. Grind on the Muller for 1×100 revolutions. Remove from the Muller plates with the palette knife and return to the nonabsorbent surface.

Perform the two steps above for the standard lot and sample lot(s). Ensure the Muller plates are cleaned well between lots. The dispersion made in these steps will be used for the masstone display and for preparation of the tint dispersion.

Procedure – Tint Preparation:

Weigh on the analytical balance the following:

- 0.200 gm Masstone dispersion from above procedure
- 4.000 gm bleaching white

Transfer to a glass plate and mix well with a palette knife until uniform in color.

Perform the above two steps for the standard lot and sample lot(s). The dispersion made in this step will be used for the tint display and for the instrumental evaluation.

Procedure – Masstone Visual Display Preparation:
Label the display sheets to include:

- Product code: top center of sheet
- Date: top right corner of sheet
- Technician initials
- Standard lot number: above the area in which the standard lot will be displayed. This position will depend on the number of lots to be displayed versus standard.
- Sample lot number: above the area in which the sample lot will be displayed.

Using the palette knife, place a small amount of the standard lot and sample lot(s), beneath their lot numbers, on the display sheet.

Hold the drawdown knife perpendicular to the display sheet surface, just above the masstone dispersions. Using even pressure, draw the knife toward the bottom of the sheet, until the dispersions have crossed the black stripe on the drawdown sheet. At this point, reduce the angle of the spatula to the sheet to approximately 45 degrees and, with reduced pressure, continue pulling the knife to the bottom of the sheet.

Using the procedure in the step immediately above, the result should be a thin film at the top of the sheet and a thicker film allowing no transmitted light at the bottom of the sheet. The dispersions should meet in a continuous flow from just below the top of the dispersion to the bottom. If there is a break between the dispersions, they do not meet, or the film thickness is not acceptable, repeat the two previous steps immediately above, varying the proximity of the initial placements and/or the pressure used to draw down the dispersions.

The display prepared in this procedure should be evaluated prior to placing a cover over the dispersion to provide the best visual conditions.

A clear plastic shield may be placed over the display to preserve for later reference; however, the dispersions may change slightly with time and the shield may distort the visual assessment.

Procedure – Tint Visual Display Preparation:
Label the display sheets to include:

- Product code: top center of sheet
- Date: top right corner of sheet
- Technician initials

- Standard lot number: above the area in which the standard lot will be displayed. This position will depend on the number of lots to be displayed versus standard.
- Sample lot number: above the area in which the sample lot will be displayed.

Using the palette knife, place a small amount of the standard lot and sample lot(s), beneath their lot numbers, on the display sheet.

Hold the drawdown knife perpendicular to the display sheet surface, just above the masstone dispersions. Using even pressure, draw the knife toward the bottom of the sheet, until the dispersions have crossed the black stripe on the drawdown sheet. At this point, reduce the angle of the spatula to the sheet to approximately 45 degrees and, with reduced pressure, continue pulling the knife to the bottom of the sheet.

Using the procedure in the step immediately above, the result should be a thin film at the top of the sheet and a thicker film allowing no transmitted light at the bottom of the sheet. The dispersions should meet in a continuous flow from just below the top of the dispersion to the bottom. If there is a break between the dispersions, they do not meet, or the film thickness is not acceptable, repeat the two steps immediately above, varying the proximity of the initial placements and/or the pressure used to draw down the dispersions.

The display prepared in this procedure should be evaluated prior to placing a cover over the dispersion to provide the best visual conditions.

A clear plastic shield may be placed over the display to preserve for later reference; however, the dispersions may change slightly with time and the shield may distort the visual assessment.

Procedure – Instrumental Display Preparation:
Place the tint dispersion on the cutout section of the template (template at this time includes backing and middle cutout).

Spread the dispersion in the cutout section of the template. Using a spatula, draw down the dispersion to a uniform thickness that fills the entire cutout. Place top cover on template to insure cleanliness of display.

Label the template with the product code and lot number. Perform the three steps above for the standard lot and sample lot(s).

Procedure – Interpretation of Results – Visual Assessment:
Visually assess the difference between the sample(s) and standard in each display, recording any differences in masstone, transparency, strength, and shade.

Procedure – Interpretation of Results – Instrumental Assessment:
Refer to the instrument operating manual for the spectrophotometric determination of the color differences.

Test 2 – Talc Evaluation

Purpose: To evaluate cosmetic pigments (organic and inorganic) in a talc dispersion to indicate performance in eye and face powder applications.

Process: The pigment is dispersed into a talc base through high speed blending in a kitchen-type blender.

Materials:

Talc (CAS # 14807-96-6)

Equipment:

Analytical balance accurate to 0.001 gm
Kitchen-type blender
8 oz glass jars
Gaskets for kitchen blender
Uncoated stock
3 1/2″ Wide drawdown knife spatula
Soft white tissue paper (e.g. rolled toilet paper)
Laboratory press capable of minimum 1000 psi
Powder die 1 1/8″ diameter
Powder pans to match the diameter of the die

Weigh into an 8 oz jar:

Pigment	0.400 gm
Talc	19.600 gm
	20.000 gm

Place two gaskets (or more if required to avoid dusting) on blender blade plate. Place plate over jar mouth, blade inserted into jar, and secure in place with blender top.

Secure jar in place on blender motor base. Blend for one minute at high speed. Remove jar from base and, by tapping the sides and bottom with palette knife, loosen packed pigment and talc from interior surfaces.

Secure jar in place on blender motor base and blend for one additional minute using the high setting on the blender.

Perform the three steps above for the standard lot and sample lot(s).

Procedure – Visual Display Preparation:
Place a full spatula tip of each dispersion prepared in the procedure above, side by side, on the uncoated stock.

Using the edge of the spatula, form the dispersions into rectangular shapes and move into position on the stock so that each dispersion just touches the next.

Cover the dispersions with tissue paper. Holding the top edge of the tissue paper in place, apply steady, even pressure from top to bottom with the drawdown knife. Remove the tissue paper from the dispersions just prior to making visual assessment.

Procedure – Instrumental Display Preparation:
Place pan in the recess of the die cylinder and fill with test dispersion. Place the die piston into cylinder and set between the plates on the hydraulic press.

Apply 1000 psi for 15 seconds using the hydraulic press.

Remove powder die from press and remove pan from bottom of the cylinder.

Perform the three steps above for standard and sample lot(s)

Procedure – Interpretation of Results – Visual Assessment:
Visually assess the difference between the sample(s) and standard in each display, recording the differences in shade and strength.

Procedure – Interpretation of Results – Instrumental Assessment:
Refer to the instrument operating manual for the spectrophotometric determination of the color differences.

Test 3 – Nitrocellulose Evaluation

Purpose: To evaluate cosmetic pigments in nitrocellulose chip dispersion for nail lacquer applications.

Process: Pigment is dispersed in nitrocellulose on a two roll mill. The chip resulting from this milling is then diluted in a solvent base to simulate a nail lacquer.

Materials:

Castor oil
1/2 sec RS Nitrocellulose (70% N/C,30% isopropanol)
Dibutyl phthalate
Butyl acetate
Ethyl acetate
White nitrocellulose paste

Equipment:

Balance accurate to 0.01 gm
Two roll mill with cooling water
$3'' \times 8''$ standard junior lab mill
Red Devil shaker
Russell 3 1/2" palette knife
Solid white Morest Chart, 7 5/8" × 10 1/4", Leneta Form WB
Morest Opacity Chart, 7 5/8" × 10 1/4", Leneta Form 5C
Vacuum plate
0.015 mm Bird Applicator
1 gal, 8 oz, and 4 oz glass bottles
400 cc beaker

Procedure – Cut Back Solution Preparation:

Weigh the following into a 1 gallon glass container (utilize appropriate measures to insure good ventilation):

Butyl acetate	400.0 gm
Ethyl acetate	600.0 gm

Seal lid on container and place on lab roller for one hour.

This mixture will accommodate approximately 14 samples and can be stored in an appropriate cabinet for flammables. A new mixture should be prepared if there is insufficient solution to complete the testing of all samples in a particular series.

Procedure – Chip Preparation:

Weigh the following into a 400 cc beaker:

Pigment	25.0 gm
1/2 sec R/S 70%	78.0 gm
Dibutyl phthalate	18.0 gm
Castor oil	2.0 gm

Mix well with palette knife.

Check the temperature of the two-roll mill prior to starting. Cooling water must be used to cool the rolls while milling. The mill gap should be tight.

Start the mill and add the mixture from the first step using the palette knife, completely emptying the beaker contents onto the mill.

When the charge first becomes a homogeneous mass and covers the front roll, start a 5 minute timer. Run the chip for 5 minutes, frequently cutting with the mill doctor blade. Monitor the chip for brittleness throughout the procedure. If the chip appears to be very dry and brittle before the time has expired, discontinue the milling.

When the time has expired, remove the chip from the mill using the doctor blade. Set the chip on a clean surface in a well-ventilated area and allow to cool.

Perform the six steps above for the standard and sample lot(s) The chip prepared in this step is:

- 100% solids
- 25% pigment

Procedure – Nail Lacquer Preparation:
Weigh the following into an 8 oz bottle:

Chip from procedure above	18.0 g
Cut-back solution	82.0 g

Perform the step above for the standard and sample lot(s).

Seal the lids on the bottles and place on the Red Devil Shaker for 45 minutes.

Remove from the shaker and allow lacquers to cool.

Procedure – Tint Preparation:
Weigh the following into a 4 oz bottle:

Lacquer from step above	5.0 g
White nitrocellulose paste	50.0 g

Perform step above for standard and sample lot(s). Seal lids on the bottles and place on Red Devil Shaker for 15 minutes.

Procedure – Masstone Visual and Instrumental Display Preparation:
Label Morest Chart SC to include:

- Product code: top center of sheet
- Date: top right corner of sheet
- Technician initials: beneath test number
- Standard lot number: above the area in which the standard lot will be displayed. This position will depend on the number of lots to be displayed versus standard.
- Sample lot number: above the area in which the sample lot will be displayed.

Place the Morest Chart on a vacuum plate to hold in place during the drawdown procedure. A sheet of paper should be placed beneath the chart at the bottom to catch any overflow of the drawdowns.

Using a pipette, place a small amount of the masstone dispersions of standard lot and sample lot(s) beneath their lot numbers on the Morest Chart.

Place the 0.015 Bird just above the dispersions. Using even pressure, draw the Bird to the bottom of the sheet.

The result should be a uniform film. The dispersions should meet in a continuous flow from just below the top of the dispersion to the bottom. If there is a break between the dispersions, they do not meet, or the film thickness is not acceptable, repeat steps above, varying the proximity of the initial placements and/or the rate used to draw down the dispersions.

Set the display aside in a clean area to dry.

The display must be completely dry prior to evaluation.

Procedure – Tint Visual and Instrumental Display Preparation:
Label Morest Chart 4B to include:

- Product code: top center of sheet
- Date: top right corner of sheet
- Technician initials: beneath test number
- Standard lot number: above the area in which the standard lot will be displayed. This position will depend on the number of lots to be displayed versus standard.
- Sample lot number: above the area in which the sample lot will be displayed.

Place the Morest Chart on a vacuum plate to hold in place during the drawdown procedure. A sheet of paper should be placed beneath the chart at the bottom to catch any overflow of the drawdowns.

Using a pipette, place a small amount of the tint dispersions of standard lot and sample lot(s) beneath their lot numbers on the Morest Chart

Place the 0.015 Bird just above the dispersions. Using even pressure, draw the Bird to the bottom of the sheet.

Using this procedure, the result should be a uniform film. The dispersions should meet in a continuous flow from just below the top of the dispersion to the bottom. If there is a break between the dispersions, they do not meet, or the film thickness is not acceptable, repeat steps above, varying the proximity of the initial placements and/or the rate used to draw down the dispersions.

Set the display aside in a clean area to dry.

The display must be completely dry prior to evaluation.

Procedure – Interpretation of Results – Visual Assessment:
Visually assess the difference between the sample(s) and standard in each display, recording any differences in masstone, transparency, strength, and shade.

Procedure – Interpretation of Results – Instrumental Assessment:
Refer to the instrument operating manual for the spectrophotometric determination of the color differences.

Test 4 – Staining Color Evaluation

Purpose: To evaluate pH sensitive fluorescein dyes (Orange 5, Red 21, Red 27) for their staining properties.

Process: Dye is ground in the presence of 1/8″ steel shot in an acrylic vehicle on a paint shaker and let down with an acrylic vehicle. Color comparisons are made versus standard run at the same time.

Materials:

Hydrofast White
Grind vehicle
Letdown vehicle

Equipment:

#8 and #12 Meyer rods
Morest Cards – coated, black-white, and white
1/8″ steel shots
8 and 4 oz glass jars
Electronic balance accurate to 0.01 gm
Harbil paint shaker

White bleach:
200 gm water
1800 gm Hydrofast White Roll on rollers for at least an hour

Procedure – Masstone Preparation:

Into an 8 oz jar, weigh the following:

1/8″ steel shots	300.0 gm
Dye	5.0 gm
Grind vehicle	95.0 gm
TOTAL	400.0 gm

Mix for one-half hour on a Harbil mixer paint shaker. This is the premix.

Add 60.0 gm of letdown vehicle and return to the Harbil shaker for 5 minutes. This is the finished masstone.

Perform the three steps above for the standard lot and sample lot(s).

Procedure – Tint Preparation:

Weigh 40.0 gm of Hydrofast White into a 4 oz jar. Rotate jar to coat all sides with bleach. Weigh in 4.0 gm of the finished masstone from steps above (1 : 10 ratio). Shake for 15 minutes on the Harbil paint shaker or until homogeneous.

Procedure – Masstone Visual Display Preparation:

Label the display sheets to include:

- Product code: top center of sheet
- Date: top right corner of sheet
- Technician initials: beneath test number
- Standard lot number: above the area in which the standard lot will be displayed. This position will depend on the number of lots to be displayed versus standard.
- Sample lot number: above the area in which the sample lot will be displayed.

Using the palette knife, place a small amount of the standard lot and sample lot(s), beneath their lot numbers, on the display sheet.

Draw down both dispersions with a number 8 Meyer rod on a black and white Morest Card. Compare by visual assessment.

Procedure – Tint Visual and Instrumental Display Preparation:
Label the display sheets to include:

- Product code: top center of sheet
- Date: top right corner of sheet
- Technician initials: beneath test number
- Standard lot number: above the area in which the standard lot will be displayed. This position will depend on the number of lots to be displayed versus standard.
- Sample lot number: above the area in which the sample lot will be displayed.

Using the palette knife, place a small amount of the standard lot and sample lot(s), beneath their lot numbers, on the display sheet.

Draw down both dispersions with a number 12 Meyer rod on a black and white Morest Card. Compare by visual or instrumental assessment.

Appendix B
Treated Pigment Patents

Patent/ application number	Description	Grant/pub. date	Assignee
US4606914	A cosmetic composition for make-up containing an N-acylamino acid salt or Al, Mg, Ca, Zn, Zi or Ti which may contain a pigment and/or an extender pigment treated with one or more of the N-acylamino acid metal salts.	Aug. 19, 1986	Miyoshi Kasei, Inc.
US4622074	Pigments and extender pigments which are surface-treated with hydrogenated lecithin, and cosmetics containing the same.	Nov. 11, 1986	Miyoshi Kasei, Inc.
US4919922	Pigment with its surface coated with a polyolefin carrying-COOR groups (wherein R is hydrogen atom or a metal atom) is water-repellent and has very high affinity for oily cosmetic components. Cosmetic products containing said surface-treated pigment is excellent in spreadability and feeling.	Apr. 24, 1990	Miyoshi Kasei, Inc.

Coloring the Cosmetic World: Using Pigments in Decorative Cosmetic Formulations, Second Edition. Edwin B. Faulkner. Edited by Jane C. Hollenberg. © 2021 John Wiley & Sons Ltd. Published 2021 by John Wiley & Sons Ltd.

Patent/ application number	Description	Grant/pub. date	Assignee
US5368639	Disclosed herein are (1) an organosilicon-treated pigment which comprises a pigment or extender pigment and a linear reactive alkylpolysiloxane having in the molecule amino groups, imino groups, halogen atoms, hydroxyl groups, or alkoxyl groups, which is adhered in an oriented mode to the surface of the pigment or extender pigment by heat treatment, (2) a process for producing said treated pigment, and (3) a cosmetic made with said treated pigment. The organosilicon-treated pigment is characterized by silicone firmly adhered to the surface of a pigment or extender pigment, freedom from residual hydrogen, very smooth feel, good adhesion to the skin, and ability to permit the color pigment of fine particle size to spread well. The treated pigment finds use as a component of high-quality cosmetics such as powder foundation, liquid foundation, rouge, and eye shadow.	Nov. 29, 1994	Miyoshi Kasei, Inc.
US7374783B2	There is provided a coated powder having a high skin care effect and a high anti-aging effect. The powder which can be used in cosmetics is coated with a mixture (lipoamino acid composition) comprising N-acyl derivatives (also including the form of a salt) of (1) at least one amino acid selected from proline and hydroxyproline, (2) at least one amino acid selected from alanine, glycine and sarcosine and (3) at least one amino acid selected from aspartic acid and glutamic acid, and at least one selected from fatty acids (and/or metal salts thereof) having a carbon number of at least 12 and at most 22.	May 20, 2008	Miyoshi Kasei, Inc.

Patent/ application number	Description	Grant/pub. date	Assignee
US5108736	Cosmetic pigments coated to be readily dispersible in oil with from 0.01 to 5.0 weight percent of the pigments of a titanate coupling agent selected from the group consisting of liquid monalkoxy (C_1-C_{20}) triisostearoyl titanates, liquid monalkoxy (C_1-C_{20}) diisostearoyl methacryl titanates, liquid monalkoxy (C_1-C_{20}) dimethacryl isostearoyl titanates, liquid monalkoxy (C_1-C_{20}) trimethacryl titanates and liquid coordinate titanates ...	Apr. 28, 1992	Kobo Products, Inc.
US20100003290Al EP2349179A1	Coated powder useful in anhydrous cosmetic composition, e.g., lipstick or pressed powder, comprises powder substrate having coating of 2-(perfiuoroalkyl) ethyl alcohol phosphate compound.	Jan. 7, 2010	Kobo Products, Inc.
US20100136065Al US8623386B2	Cosmetic composition for formulation into makeup product including lipstick, comprises pigment including red iron oxide created with jojoba ester and/or hydrogenated jojoba oil as natural surface modifying agent.	Jun. 6, 2010	Kobo Products, Inc.
EP2274054A2	Cosmetic composition for formulation into makeup product including lipstick, comprises pigment including red iron oxide treated with jojoba ester and/or hydrogenated jojoba oil as natural surface modifying agent.	Jan. 19, 2011	Kobo Products, Inc.

Patent/application number	Description	Grant/pub. date	Assignee
JP03246210A2	PURPOSE: To obtain a cosmetic having improved water repellency and oil repellency, containing powder for make-up and/or colorant subjected to surface treatment with a fluorine compound. CONSTITUTION: A cosmetic wherein at least one of extender material, other powder for make-up and a colorant to be blended as main components is subjected to surface treatment with a fluorine compound, preferably at least one compound selected from the group consisting of fluoroalkyldi(oxyethyl) amine phosphoric ester groups shown by formula I and formula II (n is 6–18, especially 8–11). The cosmetic has no make-up disorder caused by tear, rain, sweat, lipid, etc., can prevent caking phenomena, and, when the compound shown by formula I or formula II is used as the fluorine compound, suppressing effects on dust adhesion after application is obtained.	Nov. 1, 1991	Daito Kasei Kogyo KK
JP04330007A2	PURPOSE: To obtain a cosmetic having stable effect on preventing caking phenomenon by compounding a pigment surface treated with a metal hydroxide or a metal salt and further surface treated with a fluorine compound.	Nov. 18, 1992	Daito Kasei Kogyo KK

Patent/application number	Description	Grant/pub. date	Assignee
	CONSTITUTION: At least one of an extender pigment, white pigment or color pigment to be compounded into a cosmetic is surface treated with at least one kind of gel selected from hydrate, partial dehydrate and anhydride of a metal hydroxide or a metal salt and then surface treated with a fluorine compound, preferably a fluoroalkyldi(oxyethyl)amine phosphoric acid ester of formula or formula II ((n) is 6–18) (not shown). The metal hydroxide or metal salt is preferably a hydroxide or salt of magnesium, aluminum, silicon, titanium, zinc, zirconium or barium. A stable fluorine compound layer is formed by the surface treatment with the metal hydroxide or metal salt to attain a stable effect on preventing caking phenomenon.		
JP2002363444A2	PROBLEM TO BE SOLVED: To obtain a pigment for a cosmetic, capable of providing very high oil absorption and soft feeling when being formulated with a makeup cosmetic, e.g., a foundation, an eye shadow and a rouge, and having characteristics of the reflection of light being soft, and further to provide the cosmetic containing the pigment for the cosmetic. SOLUTION: The cosmetic containing the surface covered pigment for the cosmetic is prepared by formulating the surface covered pigment for the cosmetic obtained by surface covering a pigment powder for the cosmetic with 0.1–30 wt.% linear alkyl acrylate copolymer.	Dec. 18, 2002	Daito Kasei Kogyo KK

Patent/ application number	Description	Grant/pub. date	Assignee
JP2003026958A2	PROBLEM TO BE SOLVED: To provide a water dispersible pigment which can be easily dispersed in neutral water to exhibit excellent performances in moisture stability and coating stability, and an aqueous pigment dispersing solution containing the pigment dispersed in water. SOLUTION: The water dispersible pigment comprises being produced by surface modifying a pigment powder with an alkoxy silane compound represented by $[R1O(CH_2CR2HO)_n(CH_2)_k]_x Si(OC_mH_{2m+1})_y$ (wherein n and k are each an integer of 1 or more, m is an integer of 1–3, x and y are each an integer of 1–3, x + y = 4, and RI and R2 are each hydrogen or a C 1–10 hydrocarbon group). The aqueous pigment dispersing solution is produced by dispersing the pigment powder surface treated with the alkoxy silane compound in water.	Jan. 29, 2003	Daito Kasei Kogyo KK
JP2007238690A2	PROBLEM TO BE SOLVED: To provide a water and oil repellent pigment free from adverse effect on human body and having no inhibitory action on the stability of cosmetic and provide a cosmetic containing the agent. SOLUTION: The water- and oil-repellent pigment is produced by the surface treatment of a powdery pigment with tridecafluorooctyltriethoxysilane expressed by chemical formula (1). (not shown).	Mar. 7, 2006	Daito Kasei Kogyo KK

Patent/ application number	Description	Grant/pub. date	Assignee
JP2007284483A2	PROBLEM TO BE SOLVED: To provide a hygroscopic pigment capable of affording, without using collagen, a cosmetic that has a fresh bare skin feeling and a use feeling and exhibits excellent hygroscopicity, and a cosmetic comprising the same. SOLUTION: The hygroscopic pigment endowed with hygroscopicity is obtained by coating the surface of a pigment with a polysaccharide derived from a plant which lives on land. The hygroscopic pigment is incorporated with a cosmetic to allow the vegetable polysaccharide to exhibit hygroscopicity. Thus, the cosmetic that has a fresh bare skin feeling and a use feeling and exhibits excellent hygroscopicity is obtained.	Nov. 1, 2007	Daito Kasei Kogyo KK
JP2009000065901	A silane compound represented by formula (1): $(RC_nH_{2n})_aSi(OC_mH_{2m+1})_b$, a metal salt, and a pigment powder are mixed and dispersed in a mixed solvent of a C_{1-4} alcohol and water to surface treat the pigment powder with the silane compound, wherein R is a C_{3-21} hydrocarbon, a polyether group or a perfluoroalkyl group; the groups may have a functional group such as an amino group or a hydroxyl group in the chain length, have a straight chain or branched chain, and may have a single chain length or a multiple chain length; n is an integer of 1 to 12, and m is an integer of 1 to 3; and a and b are each an integer of 1 to 3, and $a + b = 4$.	Sep. 30, 2010	Daito Kasei Kogyo KK

Patent/ application number	Description	Grant/pub. date	Assignee
J P2008031138A2	The composite organic powder is characterized by mechanically coating the surface of semi-spherical to approximately semi-spherical organic powder with an inorganic pigment treated with an octyltrialkoxysilane and having a primary particle diameter of <1 µm and further coating with a colored pigment and/or an organic dye, and the cosmetic compounded with the same.	Feb. 14, 2008	Daito Kasei Kogyo KK
JP2010208986A2	PROBLEM TO BE SOLVED: To provide a surface treated pigment exhibiting water and oil repellency by virtue of a fluorine compound free of perfluorooctanoic acid (PFOA); and a cosmetic containing the pigment. SOLUTION: This pigment is obtained by surface treating a pigment powder with a fluorine compound expressed by chemical formula (not illustrated) which is a fluorine compound phosphoric acid or a salt thereof containing 100 ppb or less amount of perfluoroctanoic acid as an impurity in the compound, where in the formula, n and m are each an integer of 1 or more provided that n + m = 3; and M is a hydrogen atom or a monovalent metal ion, ammonium salt or diethanolamine salt.	Sep. 24, 2010	Daito Kasei Kogyo KK

Patent/ application number	Description	Grant/pub. date	Assignee
JP2010083846A2	PROBLEM TO BE SOLVED: To provide a cosmetic pigment which can give sufficient water resistance to a cosmetic pigment, imparts wet touch, is excellent in reactivity with a pigment, easily disperses in a hydrocarbon solvent, and is prepared into a lowly viscous formulation. SOLUTION: The surface of a pigment is coated with an alkylalkoxypolysiloxane represented by chemical structural (not shown) provided that in the formula, R_1 to R_3 are each a saturated hydrocarbon group with one or more carbons; and R_4 denotes a methoxy or ethoxy group.	Apr. 15, 2010	Daito Kasei Kogyo KK
JP2009046643A2	PROBLEM TO BE SOLVED: To produce a cellulose coated pigment that has moisture absorption and retention abilities, excellent feel in use and excellent dispersibility in an aqueous system. SOLUTION: The cellulose-coated pigment is produced by dispersing viscose solution and a pigment and then heating or using an acid to directly coat the surface of a pigment with regenerated cellulose having a crystal structure of Cellulose-II. A cosmetic material produced by blending the cellulose coated pigment into a cosmetic material is also provided.	Mar. 5, 2009	Daito Kasei Kogyo KK

Patent/ application number	Description	Grant/pub. date	Assignee
JP2008137921A2	PROBLEM TO BE SOLVED: To provide a pigment powder for a water- and oil-repellent cosmetic, obtained by subjecting a pigment powder for the cosmetic comprising a natural or artificial clay mineral to surface treatment with a perfluoroalkylalkoxysilane compound, and exhibiting sufficient water and oil repellency; and to provide the cosmetic containing the pigment powder. SOLUTION: The pigment powder for the water and oil repellent cosmetic, exhibiting the sufficient water- and oil-repellency is obtained by surface treating the pigment powder for the cosmetic with aluminum hydroxide before the surface treatment of the pigment powder for the cosmetic comprising the natural or artificial clay mineral with the perfluoroalkylalkoxysilane compound. The cosmetic containing thus obtained water and oil repellent pigment powder for the cosmetic has excellent water and oil repellency and good durability of the makeup.	Jun. 19, 2008	Daito Kasei Kogyo KK
JP2008050387A2	PROBLEM TO BE SOLVED: To obtain a water-repellent and oil-repellent pigment which provides a cosmetic with water repellency and oil repellency when mixed with the cosmetic, has a fitness feeling to the skin and no feel of powder likeness, to provide a method for producing the same and to obtain a cosmetic containing the same.	Mar. 6, 2008	Daito Kasei Kogyo KK

Patent/ application number	Description	Grant/pub. date	Assignee
	SOLUTION: The water-repellent and oil-repellent pigment subjected to surface treatment with a perfluoroacrylate copolymer represented by chemical structural formula (1) is obtained through a process for dissolving the perfluoroacrylate copolymer in an organic solvent and mixing the solution with pigment powder by a mixing disperser. The cosmetic mixed with the water repellent and oil repellent pigment controls the condition of makeup fading due to perspiration and dullness or color and has fitness feeling to the skin and no feel of powder likeness.		
JP2007326902A2	PROBLEM TO BE SOLVED: To provide a pigment for cosmetic materials with which a high concentration powder can be dispersed in any oil agent for cosmetic materials, and a cosmetic material containing the pigment for cosmetic materials. SOLUTION: The pigment for cosmetic materials is provided by coating the powdery pigment for cosmetic materials with a specific polysiloxane compound, e.g., a hydroxymethyl polysiloxane, a hydroxyethyl polysiloxane, a hydroxyphenyl polysiloxane, subsequently reacting the coated pigment with a specific alkylalkoxysilane in water.	Dec. 20, 2007	Daito Kasei Kogyo KK

Patent/ application number	Description	Grant/pub. date	Assignee
JP2007070500A2	PROBLEM TO BE SOLVED: To provide a method of manufacturing water dispersible pigment which can give water dispersibility regardless of the kinds of pigments, and is superior in long time water dispersibility and also in coating stability, and to provide a water dispersion of the water dispersible pigment. SOLUTION: This water dispersible pigment is obtained by applying an oxygen plasma treatment or an ammonia plasma treatment to a surface-reformed pigment powder obtained by coating the surface of a pigment powder with an organopolysiloxane then heat treating it. Its water dispersion is obtained by dispersing the thus obtained water dispersible pigment in water.	Mar. 23, 2007	Daito Kasei Kogyo KK
JP2003160441A2	PROBLEM TO BE SOLVED: To provide a moisture absorbing and retaining pigment having an excellent feeling when used, and excellent in moisture absorption and retention, without using collagen, and to provide a cosmetic compounded with the pigment, having an excellent moisture feeling in an area to which the cosmetic is applied, and capable of keeping the moisture feeling.	Jun. 3, 2003	Daito Kasei Kogyo KK

Patent/ application number	Description	Grant/pub. date	Assignee
	SOLUTION: This moisture absorbing and retaining pigment is obtained by subjecting surfaces of the pigment to surface coating treatment with a synthetic or semi-synthetic water soluble polymer. The pigment has the excellent feeling when used and is excellent in moisture absorption and retention. Further, the cosmetic is compounded with the moisture absorbing and retaining pigment. The cosmetic has the excellent moisture feeling in the area to which the cosmetic is applied, and further keeps the moisture feeling.		
US20110110995Al	Provided is a surface-created powder having excellent usability and adhesion to skin. The powder is coated with a surface treating agent including (a) a fluorine-containing monomer of a general formula (I) and (b) an alkoxy group containing monomer of a general formula (II). The powder is used for various cosmetics. [Chemical formula 3] $CH_2 = C(-X)-C(PO)-Y-[-(CH_2)_m-Z-]_p-(CH_2)_n-R_f$ (I) [Chemical formula 4) $CH_2 = C(Xl)-C(PO)-O-(RO)q-X2$ (11).	May 12, 2011	Miyoshi Kasei, Inc.

Patent/ application number	Description	Grant/pub. date	Assignee
EP2289484AJ	Provided is a surface-treated powder having excellent usability and adhesion to skin. The powder is coated by copolymerizing surface-treating agents including (a) a fluorine-containing monomer of a general formula (I) and (b) an alkoxy group-containing monomer of a general formula (II). The powder is used for various cosmetics. I: $CH_2=C(-X)-C(=O)-Y-[-(CH_2)_m-Z-]_p-(CH_2)_n-R_fX=H$, CH_2, Cl, Br, I, a CFX1X2 in which X1 and X2 are a hydrogen atom, a fluorine atom or a chlorine atom, a cyano group, a straight chain or branched-chain fluoroalkyl group having 1 to 20 carbon atoms, or a substituted or non-substituted benzyl group, a substituted or non-substituted phenyl group; Y is -O- or -NH-; Z is a direct bond, -S- or -SO2-; R_f is a fluoroalkyl group having 1 to 6 carbon atoms; m is 1 to 10, n is 0 to 10, and p is 0 or 1. II: $CH_2=C(X1)-C(=O)-O-(RO)_q-X2$.	Mar. 2, 2011	Miyoshi Kasei, Inc.; Daikin Industries, Ltd.
US6790452	A pigment containing particles surface created with N-acylglycine hydroxyaluminum and/or N-acyl-N-methylglycine hydroxyaluminum, and a pigment dispersion which contains such surface treated particles and a cosmetically acceptable oily material, having excellent dispersion stability and versatility (case of dispersion and wide range of usage), because the particles are easily and uniformly dispersed in a short period of time at a high concentration in such a cosmetically acceptable oily material, thus a cosmetic product is thereby stable as an emulsion and has better usage than existing products.	Sep. 14, 2004	Miyoshi Kasei, Inc.

Patent/application number	Description	Grant/pub. date	Assignee
US6296860	Pigments and extender pigments coated with a mixture comprising at least N-acyl forms of naturally existing 14 amino acids, in particular, pigments and extender pigments processed with a mixture comprising at least N-acyl forms of 14 amino acids, obtainable from or obtained on hydrolysis of any one of animal proteins, such as silk and pearl, and plant proteins, such as wheat and soybeans, which can be mixed into cosmetics, and also cosmetics comprising the coated pigments and extender pigments are provided. Such N-acyl form therein may be in the salt form. The obtained pigments and extender pigments coated therewith in the present invention has extremely smooth tactile feeling that is an optimum affinity to skin and/or hair (biocompatibility) when used in cosmetics, excellent application feeling on skin and/or hair without coarse feeling, and hence suitable to be combined in cosmetic powders.	Oct. 2, 2001	Miyoshi Kasei, Inc.

Patent/ application number	Description	Grant/pub. date	Assignee
US5458681	A pigment or extender pigment created with a linear reactive alkylpolysiloxane having in the molecule amino groups, imino groups, halogen atoms, hydroxyl groups, or alkoxyl groups, which is oriented and adsorbed to the surface of the pigment or extender pigment by heat treatment; a process for producing the treated pigment; and a cosmetic made with the treated pigment. The alkylpolysiloxane has a degree of polymerization from 25 to 100 and a Mw/Mn ratio from 1.0 to 1.3. The organosilicone treated pigment, characterized by silicone firmly adsorbed to its surface, freedom from residual hydrogen, very smooth feel, good adhesion to the skin, and ability to permit color pigment of fine particle size to spread well, is particularly suitable for use in cosmetics such as powder foundation, liquid foundation, rouge, and eye shadow.	Oct. 17, 2005	Miyoshi Kasei, Inc.
US5368639	Disclosed herein are (1) an organosilicon treated pigment which comprises a pigment or extender pigment and a linear reactive alkylpolysiloxane having in the molecule amino groups, imino groups, halogen atoms, hydroxyl groups, or alkoxyl groups, which is adhered in an oriented mode to the surface of the pigment or extender pigment by heat treatment, (2) a process for producing said treated pigment, and (3) a cosmetic made with said created pigment. The organosilicon treated pigment is characterized by silicone firmly adhered to the surface of a pigment or extender pigment, freedom from residual hydrogen, very smooth feel, good adhesion to the skin, and ability to permit the color pigment of line particle size to spread well. The treated pigment finds use as a component of high quality cosmetics such as powder foundation, liquid foundation, and rouge.	Nov. 29, 1994	Miyoshi Kasei, Inc.

Patent/ application number	Description	Grant/pub. date	Assignee
US5091013	A pigment (or body) composed of particles having surfaces covered with a high molecular substance containing a group of the formula --COOR, where R stands for a hydrogen or metal atom, and having an acid value of at least 200 when the R in the formula stands for a hydrogen atom. It has a high power of absorbing and holding moisture. The surfaces of the pigment particles arc alternatively covered with both such a high molecular substance and a substance which makes the surfaces hydrophobic. This pigment has a good affinity for oil and yet a high power of absorbing and holding moisture. A cosmetic containing any such pigment has a high moisturizing power and provides a long makeup life.	Feb. 25, 1992	Miyoshi Kasei, Inc.
JP2011088850A2	PROBLEM TO BE SOLVED: To provide a surface treated powder which cannot be easily eluted with water and a solvent such as ethanol.	May 6, 2011	Miyoshi Kasei, Inc.

Patent/ application number	Description	Grant/pub. date	Assignee
	SOLUTION: The surface treated powder is obtained by coating the surface of the powder particle To be surface treated with a specific surface treating agent and the surface treating agent contains a fluoroalkyl acrylate/polyalkylene glycol acrylate polymer obtained by copolymerizing a monomer including (a) a fluorine-containing monomer represented by formula (I), $CH_2=C(-X)-C(=O)-Y-[-(CH_2)_m-Z-)p-(CH_2)_n-R_f$, and (b) an alkoxy group-containing monomer represented by formula (II), $CH_2=C(X3)-C(=O)-O-(RO)_q-X4$, as essential components, and a treating auxiliary of (c) a polyvalent acidic substance or a polyvalent basic substance as essential components. In the formulas, each symbol has a special definition.		
EP0522916B1	The present invention relates to an organosilicon treated pigment, a process for production thereof, and a cosmetic made therewith, and more particularly, to a new pigment and extender pigment which arc smooth, superior in adhesion and spreadability, and completely free of residual hydrogen, a process for production of said pigment, and a cosmetic made with said pigment. (Effect) The treated pigment of the present invention is characterized by silicon firmly adsorbed to the surface of a pigment or extender pigment, no residual hydrogen and very smooth feel, good adhesion to the skin, and ability to permit the color pigment of fine particle to spread well. Therefore, the treated pigment of the present invention finds use as a component of cosmetics such as powder foundation, liquid foundation, rouge and eye shadow and remarkably contributes to the improvement of quality of such cosmetics.	Sep. 10, 1997	Miyoshi Kasei, Inc.

Patent/ application number	Description	Grant/pub. date	Assignee
JP07196946A2	PURPOSE: To obtain a pigment treated with an organic silicon compound, having high silicone oil absorption, free from residual hydrogen, exhibiting extremely high smoothness and good adhesion, giving fine particles of colored pigment having excellent color developability and, accordingly, useful to be compounded into make-up cosmetics such as powder foundation, liquid foundation, rouge, and eye shadow to remarkably contribute to quality improvement and give a cosmetic exhibiting smooth feeling unattainable by conventional cosmetics. CONSTITUTION: This pigment or extender pigment treated with an organic silicon compound is produced by adsorbing a straight-chain reactive alkylpolysiloxane having an amino-hydrogen group, halogen atom, hydroxyl group, or alkoxy group on one molecular chain terminal and having a degree of polymerizarion of 25–100 and a Mw/Mn ratio of 1.0–1.3 (Mw is weight-average molecular weight and Mn is number-average molecular weight) to the surface of a pigment or an extender pigment by heat treatment. This invention also relates to the process for the production of the treated pigment.	Aug. 1, 1995	Miyoshi Kasei, Inc.

Patent/ application number	Description	Grant/pub. date	Assignee
JP05339518A2	PURPOSE: To obtain the title pigment which is smooth, excellent in adhesion and spreadability, and completely free from residual hydrogen by heating and drying a mixture of a specific reactive alkylpolysiloxane, an organic solvent, and a coloring or extender pigment. CONSTITUTION: A linear reactive alkylpolysiloxane having an amine hydrogen, a halogen, hydroxy, or an alkoxy in the molecule is mixed with an organic solvent solubilizing the polysiloxane and with a coloring or extender pigment (a). This mixture is dried by heating to obtain the title pigment. Since the heat treatment enhances the adsorption of the silicone to the pigment (a), the fine treated coloring pigment has excellent spreadability. The title pigment hence significantly contributes to an improvement in quality when incorporated into makeup preparations such as powder foundation, liquid foundation, rouge, eyeshadow, etc.	Dec. 21, 1993	Miyoshi Kasei, Inc.

Patent/ application number	Description	Grant/pub. date	Assignee
EP0522916A3	The present invention relates to an organosilicon treated pigment, a process for production thereof, and a cosmetic made therewith, and more particularly to a new pigment and extender pigment which are smooth, superior in adhesion and spreadability, and completely free of residual hydrogen, a process for production of said pigment, and a cosmetic made with said pigment. (Effect) The treated pigment of the present invention is characterized by silicon firmly adsorbed to the surface of a pigment or extender pigment, no residual hydrogen and very smooth feel, good adhesion to the skin, and ability to permit the color pigment of fine particle to spread well. Therefore, the treated pigment of the present invention finds use as a component of cosmetics such as powder foundation, liquid foundation, rouge and eye shadow and remarkably contributes to the improvement of quality of such cosmetics.	Jan. 27, 1993	Miyoshi Kasei, Inc.
US7964178	Modified colorants are made from a modifying agent, such as platelet alumina, and a colorant. The colorant may be fixed to the surface of the modifying agent, optionally with a surface treatment. The colorant may not completely coat the surface of the modifying agent, and the edge of the modifying agent may be substantially free of colorant. The modified colorants may be used in cosmetic products.	Jun. 21, 2011	Sensient Colors, Inc.

Patent/ application number	Description	Grant/pub. date	Assignee
US20090185984A1	Compositions including colorants surface-treated with silicone polyurethanes, and methods for making surface treated colorant products, are provided. Cosmetic, personal care, skin care, nail and hair products including the surface created colorants are disclosed. Also provided are methods for improving the hydrophobicity, or the adherence to the surface of skin, of a colorant. The methods include surface treating a colorant with an effective amount of a silicone polyurethane to form a surface treated colorant having an improved hydrophobicity or an improved adherence to skin.	Jun. 23, 2009	Sensient Colors, Inc.
US20100105817A1 US9029586B2 Silanes with embedded hydrophilicity, dispersible particles derived therefrom and related methods	The invention provides a silane compound that includes a hydrophobic group and a silane ester group linked by a hydrophilic group for use as a surface treatment to an inorganic material, such as a pigment, the silane including a hydrophobic group and a silane ester group linked by a hydrophilic group. The invention includes a coated particle including an inorganic material coated with the silane compound(s) and methods of improving the wettability and/or dispersibility of an inorganic material such as a pigment, wherein the method comprises depositing the silane compounds on the surface of a pigment.	Apr. 29, 2010 May 12, 2015	Gelest Technolo-gies, Inc.

Patent/ application number	Description	Grant/pub. date	Assignee
US9358200B2 Alkoxysilane derivatives of N-acyl amino acids, N-acyl dipeptides, and N-acyl tripeptides, and particles and stable oil-in-water formulations using the same	Hydrophilic N-acylamino acid, N-acyl dipeptide, and N-acyl tripeptide substituted silanes are prepared which can be utilized as reactive surface treatments for particles of pigments, minerals, and fillers. These treated particles form stable dispersions in the aqueous phase of oil-in-water mixtures that are suitable for cosmetic applications. The treated particles may also be used in pressed powder and color cosmetic formulations.	Jun. 7, 2016	Gelest Technologies, Inc.

Glossary of Terms

Absorbtion colorant A colorant that produces its color by selectively absorbing and reflecting different wavelengths of light.

Angle of incidence The angle at which light strikes a substance.

Angle of reflection The angle at which light is reflected from the surface of a substance.

Annex IV The table of color additives approved for use in the European Union via the Cosmetics Directive.

Bleed The slight solubility of a pigment in a vehicle.

Bottle tone The hue and strength of a finished cosmetic product as it is viewed in a container on a retail shelf.

Brightness The lightness or darkness of a color; one of Munsell's attributes of color.

Bulk An in-process cosmetic that has been processed with all its ingredients, but has not been put into a finished package (e.g., lipstick tube, pressed powder pan, liquid makeup bottle, etc.).

Chroma A term synonymous with "intensity."

Coloring the Cosmetic World: Using Pigments in Decorative Cosmetic Formulations,
Second Edition. Edwin B. Faulkner. Edited by Jane C. Hollenberg.
© 2021 John Wiley & Sons Ltd. Published 2021 by John Wiley & Sons Ltd.

Color additive   A term synonymous with "colorant," the legal term used in US Food and Drug Administration (FDA) regulations to refer to a material used to provide color to a cosmetic product or the human body.

Colorant   A chemical that is used to provide color.

Color cosmetics   Cosmetic products that are primarily designed to enhance a consumer's appearance by adding color to the skin, either to cover flaws in the complexion or to bring bright color to parts of the body (e.g., the lips, nails, and eyes). A term synonymous with "decorative cosmetics."

Color point   The point in three-dimensional color space where a particular color resides. It is the mathematical integration of the color's three attributes: hue, brightness, and intensity.

Color travel   The ability of a pigment to provide multiple color shades to a substrate as it is viewed from multiple angles.

Composite pigments   Pigments of two or more layers producing effects not seen in the individual component pigments or substrates.

Compressibility   In pressed powder products, the facility with which the ingredients in a formula can be compressed into a pan such that the resulting product is able to withstand the rigors of frequent use without breaking.

Coverage   The ability of a pigment to hide the substrate on which it is applied. Used frequently in reference to effect pigments. With absorption pigments, the synonymous term "opacity" is often used.

Decorative cosmetics   Cosmetic products that are primarily designed to enhance a consumer's appearance by adding color to the skin, either to cover flaws in the complexion or to bring bright color to parts of the body (e.g., the lips, nails, and eyes). A term synonymous with "color cosmetics."

Dichroism   The ability of a colorant to exhibit different color shades when viewed from two different angles. This is called a "dichroic effect."

Diffuse reflection   A reflection in which the angle of reflection is not equal to the angle of incidence. Where all the reflection is diffuse, it is not possible for the human eye to see any image reflected in a substance's surface.

Disperser   A term synonymous with "mill."

Dispersibilty   The degree of ease with which the aggregates and agglomerates of a particular pigment type can be separated.

Dispersion (1) The process of converting a "raw" pigment into a usable form, providing the best color and money value. (2) The process of wetting, separating, and distributing pigment particles in a vehicle. (3) The mixture of colorant and vehicle that results from the dispersion process.

Display A visual representation of the colorant dispersed in a thin film of a vehicle such as castor oil, nitrocellulose, or talc. The film is usually shown on a card with one black half and one white half.

Drawdown (1) A term synonymous with "display." (2) The process of producing a thin film of a colorant dispersion in order to make such a display.

Drawdown knife A device used to produce films of an oil dispersion (e.g., the Appendix A Castor Oil Test). Similar to a common putty knife.

Drawdown rod A device used to produce films in the Appendix A Nitrocellulose and Staining Color Tests. Commonly called a Meyer rod.

Dye A color that is soluble in the medium in which it is applied.

Effect pigment A colorant that produces its color by selectively reflecting, refracting, absorbing, and transmitting light.

Finished product In the context of this book, a finished cosmetic product such as a lipstick, nail polish, eye shadow, or liquid makeup.

Fugitive A reference to light stability. A colorant is said to be fugitive if it has very poor light stability.

Gelling A phenomenon whereby an increase in viscosity causes a product to take on a jellylike form, impacting its spreadability.

Grinder A term synonymous with "mill."

Hegman gauge A device used to measure the particle size of a pigment of a dispersion made on a mill.

Hue The shade or color of a pigment; one of Munsell's attributes of color.

Hydrophilic Of chemicals, the quality of having an affinity for water, and so attracting it.

Hydrophobic Of chemicals, the quality of not having an affinity for water, and so repelling it.

Incident light The light, composed of multiple wavelengths of the electromagnetic spectrum, that strikes a substance.

Inorganic pigment A pigment that does not contain carbon.

Intensity The measure of color strength; one of Munsell's attributes of color.

Iron Blue   Another name for the pigment ferric ammonium ferrocyanide.

Jetness   A qualitative description of the degree of blackness of a black pigment.

Light absorption   The degree to which the wavelengths of light striking a substance pass into it.

Light reflection   The degree to which the wavelengths of light striking a substance bounce back from its surface.

Light refraction   The degree to which light slows and bends as it is transmitted through a substance.

Light transmission   The ability of a substance to allow light to pass completely through it rather than being absorbed or reflected.

Lipophilic   Of chemicals, the quality of having an affinity for oil, and so attracting it.

Lipophobic   Of chemicals, the quality of not having an affinity for oil, and so repelling it.

Luster   The ability of an effect pigment to impart shine, glitter, sparkle, gloss, radiance, and sheen by reflecting light from multiple surfaces.

Masstone   The ability of a colorant to impart a deep rich tone in cosmetics. Most important in dark shades of lipstick and nail polish.

Metallic pigment   A colorant that is a pure metal, including copper, bronze, aluminum, silver, and gold.

Metamerism   The ability of a colorant to exhibit different color shades when illuminated under different types of light (e.g., incandescent vs. fluorescent). This is called a metameric effect.

Mill   The piece of equipment used to disperse a pigment. The term applies to both dry and wet dispersions.

Monochromatic dispersion   A dispersion that contains only one type of pigment.

Morest Card   The half white/half black card used to make displays of colorant.

Muller   The device used to disperse pigments in a castor oil test. Commonly referred to as a Hoover Muller, Hoover being the US manufacturer.

Muscovite Mica   A type of natural mica that is a potassium aluminum silicate.

Neat   A pure colorant, one not blended with other colorants or diluted with any extender.

Oil absorption   The measure of a pigment's ability to absorb oil.

Opacity   The property of a colorant that permits light and/or images to be seen through a color film.

Organic pigment   A pigment based on carbon chemistry.

Panelist   A person who volunteers to apply cosmetic products to their skin or to be examined by an independent evaluator to provide a qualitative or quantitative report on product performance.

Paste   A similar term to dispersion, but used exclusively when referring to pearl pigments.

Pay-off   The ease with which a cosmetic product can be applied to the skin.

Phlogopite Mica   A type of natural mica that is a potassium magnesium aluminum silicate.

Pigment   A colorant that is not soluble in the medium in which it is applied.

Pigment load   The amount of pigment, expressed as a percentage, in a given formulation.

Polychromatic dispersion   A dispersion that contains more than one type of pigment.

Precipitant   A cation that is used to precipitate a water soluble dye from solution as a metal salt (e.g., Ca^{+2}, Ba^{+2}, Na^{+}, etc.).

Rheology   The measure of a pigment's ability to affect the viscosity and flow properties of liquid cosmetics. This attribute is important in nail polish and liquid makeup.

Saturation   A term synonymous with "intensity."

SEM   An abbreviation for "scanning election microscope."

Shade   Synonymous with "hue," appearance characterized by specific wavelengths of light (e.g., blue shade red, green shade blue, etc.).

Silver   When used in the context of an effect pigment, sometimes one that does not exhibit any of the individual visible spectrum colors that are also referred to as "white."

Spatula   A knifelike, flat-bladed instrument used to weigh or mix dry ingredients into a liquid vehicle on a glass plate or glass slab.

Spectrophotometer   An instrument used to measure light intensity.

Specular reflection   A reflection in which the angle of reflection is equal to the angle of incidence. The best example is the reflection created by a mirror. The sharpness of the image seen in the mirror is directly related to the percentage of specular reflection the mirror is able to produce.

Staining properties   The ability of a dye to stain the skin.

Standard   In color measurement, a particular color against which all samples are tested and comparisons made.

Stir-in pigment   Pigments that do not require any type of high shear dispersion. They can be incorporated into a decorative cosmetic product with simple mixing.

Streaking   A separation of color when a finished cosmetic product is applied to skin, resulting in a variegated appearance.

Strength   A term synonymous with "intensity."

Surface tension   The property of a liquid caused by cohesion of its molecules that allows it to resist spreading and to support denser objects on its surface.

Table 6   The table of color additives approved for use in China via the Hygienic Standards for Cosmetics.

Tinctorial value   The degree of "intensity" a colorant can impart to a vehicle.

Tint   A display of a colorant where the original vehicle/colorant dispersion material used to make the masstone has been diluted with a white pigment, typically titanium dioxide or zinc oxide.

Transparency   The property of a colorant that permits light and images to be transmitted from a substrate back through a color film, or light to pass through a finished toiletry product.

Value   A term synonymous with "brightness" or "lightness."

Vehicle   The medium into which a pigment is dispersed (e.g., castor oil, water, nitrocellulose, talc).

Vividness   A term synonymous with "intensity."

Wavelength   The distance between successive crests of a wave of light, corresponding to a distinct color.

White   When used in the context of an effect pigment, one that does not exhibit any of the individual visible spectrum colors. "White" effect pigments may also be referred to as "silver."

Further Reading

Commission Internationale de l'Eclairage, CIE publications – premium source for knowledge on light and lighting. `http://cie.co.at/publications` (accessed September 28, 2020).

Emerton, V. (2008). *Food Colours*, 2e. Hoboken, NJ: Wiley-Blackwell.

European Union (2009). Regulation (EC) No. 1223/2009 of the European Parliament and of the Council of 30 Nov 2009, Annex IV, Part I. *Official Journal of the European Communities.* `http://eur-lex.europa.eu/LexUriServ/LexUriServ.do?uri=OJ:L:2009:342:0059:0209:en:PDF` (accessed September 28, 2020).

FDA (2018) Code of Federal Regulations Title 21, Parts 1–99, revised as of April 1, 2018. `https://www.accessdata.fda.gov/scripts/cdrh/cfdocs/cfcfr/CFRSearch.cfm?CFRPart=1-91` (accessed September 28, 2020).

Hochheiser, H.S. (1982). *Synthetic Food Colors in the United States: A History Under Regulation.* Ann Arbor, MI: University Microfilms International.

Konica Minolta (n.d.). Precise color communication. `https://www.konicaminolta.com/instruments/knowledge/color/index.html` (accessed September 28, 2020).

Kuehni, R.G. (1997). *Color: An Introduction to Practice and Principles.* Hoboken, NJ: Wiley.

Lewis, P.A. (ed.) (1988). *Pigment Handbook*, vol. 1, 2. Hoboken, NJ: Wiley.

Marmion, D. (1991). *Handbook of US. Colorants for Foods, Drugs, and Cosmetics*, 3e. Hoboken, NJ: Wiley.

Coloring the Cosmetic World: Using Pigments in Decorative Cosmetic Formulations,
Second Edition. Edwin B. Faulkner. Edited by Jane C. Hollenberg.
© 2021 John Wiley & Sons Ltd. Published 2021 by John Wiley & Sons Ltd.

Nikitakis, J. and Lange, B. (eds.) (2016). *International Cosmetic Dictionary and Handbook*, 16e. Washington, DC: The Personal Care Products Council.

Patton, T.C. (1973). *Pigment Handbook*, vol. II, III. Hoboken, NJ: Wiley.

Pfaff, G. (2009). Special effect pigments. In: *High Performance Pigments*, 2e (eds. E. Faulkner and R.J. Schwartz). Hoboken, NJ: Wiley Interscience.

Schlossman, M.L. (2002). *The Chemistry and Manufacture of Cosmetics, Volume III: Ingredients*, 3e. Carol Stream, IL: Allured Publishing.

Society of Dyers and Colourists (n.d.). The colour index. `https://colour-index.com/` (accessed September 28, 2020).

X-Rite (n.d.). A guide to understanding color. `https://www.xrite.com/-/media/xrite/files/whitepaper_pdfs/l10-001_a_guide_to_understanding_color_communication/l10-001_understand_color_en.pdf` (accessed September 28, 2020).

Yakuji, N. (1984). *Principles of Cosmetic Licensing in Japan*. Tokyo: Yakuji Nippo.

Zuckerman, S. and Senackerib, J. (2007). Colorants for foods, drugs, and cosmetics. In: *Kirk-Othmer Food & Feed Technology*, vol. 1. Hoboken, NJ: Wiley.

Index

Printed and bound by CPI Group (UK) Ltd, Croydon, CR0 4YY

16/04/2025

14658552-0001